MECHANICAL ENGINEERING PRIMER

Robert Tata, B.S.M.E., P.E.

AuthorHouse™ LLC
1663 Liberty Drive
Bloomington, IN 47403
www.authorhouse.com
Phone: 1-800-839-8640

Published by AuthorHouse 10/14/2013

ISBN: 978-1-4918-2648-5 (sc)
ISBN: 978-1-4918-2647-8 (e)

Library of Congress Control Number: 2013918396

TABLE OF CONTENTS

OVERVIEW

This book is written for the young who may want to prepare for a technical future or any other individual who may want to broaden their horizons. It is written in an easy to understand step-by-step style and contains more pages of illustrative examples than pages of text, enabling the reader to better understand the subject matter. At the end is a thirty-three question quiz should this book be used for class room study or for the challenge and enjoyment of other readers.

Engineering is a field of endeavor that includes a wide range of topics that merit attention. This course begins by dealing with some of the fundamental issues such as engineering materials, drawings, fasteners, couplings, belts and pulleys. It then provides more in depth discussion on the design of gears, bearings, shafts and how simple beam formula can be used to calculate gearbox shaft loads and deflections. It analyzes power flow through an automotive drive train from the engine, through the torque converter, transmission, drive shaft and to the wheels. It concludes with information on how to apply for an engineering patent and how a bill recently passed in Congress has changed an important aspect of the patent law.

ENGINEERING MATERIALS

Iron & Steel: Two of the most important engineering materials used today are iron and steel. Iron is the fourth most abundant element found in the earth's crust by weight following oxygen, silicon, and aluminum respectively and occurs as an ore in the form of iron oxide. Iron ore is loaded in a furnace with coke and limestone and blasted from the bottom with hot air. The coke and hot air combine to reduce the iron oxide to iron while the limestone removes impurities. This product called pig iron is further processed to make castings, pipe, and sheet stock that are used in the many industrial products that surround us today.

Most pig iron is put into huge furnaces with a small percentage of alloying elements to produce steel. Steel is the world's most important metal and is found in everything from bolts to bridges. The hardness and strength of steel can vary greatly depending on the kind and amount of alloying elements that are added to the pig iron. When an increasing amount of carbon is added to steel, its properties of hardness and strength increase. More importantly, it becomes increasingly responsive to heat treatment producing even higher hardness and strength properties. The carbon content of steel ranges from .05% to 1.0%. This effect that carbon has on hardening steel ends approximately with a maximum content of about 1.0%. Amounts of carbon over 1.7% limit the ability of steel to be responsive to hot and cold treatment and revert it back to pig iron type material.

Other important alloying elements of steel are manganese, phosphorus, sulfur, molybdenum, chromium, and nickel. Manganese is next to carbon in importance in steel additives. Manganese is usually present in quantities ranging from .5 to 2.0%. Manganese imparts strength and responsiveness to heat treatment to steel. Small quantities of phosphorus are present in all steel. Phosphorus increases the strength of steel and ductility at low temperatures. Sulfur is an important element in steel because it increases machinability. Molybdenum increases toughness and tends to resist softening at high temperatures. Chromium in the quantity of .5 to 1.5% increases response to heat treatment. Chromium in the quantity of 12 to 25%, in combination with up to 20%

nickel, increase resistance to oxidation and corrosion in what is called stainless steel. Nickel, in the quantity of 1 to 4%, increases the strength and toughness of steel.

Aluminum: Another important material used today is aluminum. Aluminum is the third most abundant element found in the earth's crust behind oxygen and silicon. Although aluminum isn't as strong as iron and steel, it is used extensively in industry because it is lighter, easier cast, and has more corrosion resistance than iron and steel. Because of the strong chemical bond between aluminum and the oxygen in its ore, it cannot be processed in a furnace like iron. Aluminum is obtained from aluminum oxide using an electrolytic process. The oxide is placed in an electrolytic cell with cryolite. The resulting reaction reduces the oxygen in aluminum oxide to carbon monoxide and carbon dioxide leaving the aluminum to settle to the bottom where it is removed and sent to a holding furnace.

For sand casting, the primary alloying elements for aluminum are copper and silicon. A general purpose casting alloy contains 8% copper. Silicon alloys of 5% are used because of their excellent casting and resistance to corrosion properties. Aluminum wrought alloys are classified into different groups: those hardened and strengthened by cold working and those strengthened by heat treatment. Pure aluminum with no alloying elements is work hardenable and is an excellent conductor of electricity. Pure aluminum with 1.25% manganese added is similar but is stronger with a little less electrical conductivity. Both of these alloys are available in a wide range of products such as sheet, rod, tube, wire, and extruded shapes. Where high strength is required, it is necessary to use heat treatable aluminum alloys. One such alloy is 2024. It is a heat-treatable alloy containing primarily 4.4% copper. Its heat treatable hardness is among the highest of all aluminum alloys and is used in the aircraft industry. Another popular heat treatable aluminum alloy used for many industrial purposes is 6061. This alloy contains small quantities of copper, silicon, magnesium, and chrome.

<u>Magnesium</u>: Magnesium is the seventh most abundant element found in the earth's crust following oxygen, silicon, aluminum, iron, calcium, and sodium. Magnesium is the third most structural metal used today behind iron and aluminum. Magnesium is obtained from seawater using electrolysis methods. Magnesium is alloyed with aluminum, zinc, and manganese and can be heat treated.

Magnesium is the lightest of the structural metals used today. It is used in the aircraft and missile field and for other industrial products where weight is of prime importance. Magnesium is available in castings, forgings, extrusions, and sheet. Magnesium is easily machined, can be riveted, welded, and has excellent corrosion resistant properties.

Material Testing: An important test in which many important strength characteristics of engineering materials can be determined is called the tensile test. The tensile test machine clamps one end of a round test bar and slowly applies a tensile load to the other end until the bar breaks. Figure 1 is an example of a tensile test plot for low carbon steel. The vertical axis is the stress applied to the test specimen in pounds per square inch (psi) and the horizontal axis is the resulting strain of the test specimen in inches per inch. The initial segment of the plot is a straight line representing the elastic region in which the test specimen will return to its original shape should the load be removed. The slope of the curve in this region is called the modulus of elasticity and is a measure of the stiffness of the material. In general, the point on the curve where the plot starts to deviate from a straight line is called the elastic limit and represents the yield strength of the material. Beyond the elastic limit, the test sample will not return to its original shape. If the load is increased further, the curve will deviate more and more from a straight line, reach a peak, and then start to descend until the test specimen ruptures. The ultimate or tensile strength is the stress that is being applied to the test specimen at the peak of the curve. The stress applied when the specimen breaks is called the rupture stress.

Material Hardness: A correlation exists between the hardness and the tensile strength of a material. Hardness is a very easy characteristic to test for and therefore has considerable use in industry. The Brinell machine is one tool that is widely used in industry to measure hardness. It is non-destructive and can be used for a wide variety of metals. It relies on the depth of penetration of a ball under load to determine the hardness of the test specimen. The following chart lists the yield strength, tensile strength, and hardness of some of the metals discussed above:

Metal	Yield Strength	Tensile Strength	Hardness
Magnesium Alloy	11,000 (psi)	21,000 (in/in)	47 (BHN)
Magnesium Alloy	30,000	45,000	48
Aluminum Alloy	34,000	34,000	100
Aluminum Alloy	40,000	48,000	105
Wrought Iron	30,000	50,000	100
Cast Iron	24,000	39,000	200
Structural Steel	35,000	57,000	120
SAE 1300 Steel	40,000	70,000	150
SAE 4300 Steel	45,000	80,000	170

Figure 1

Stress-Strain Diagram

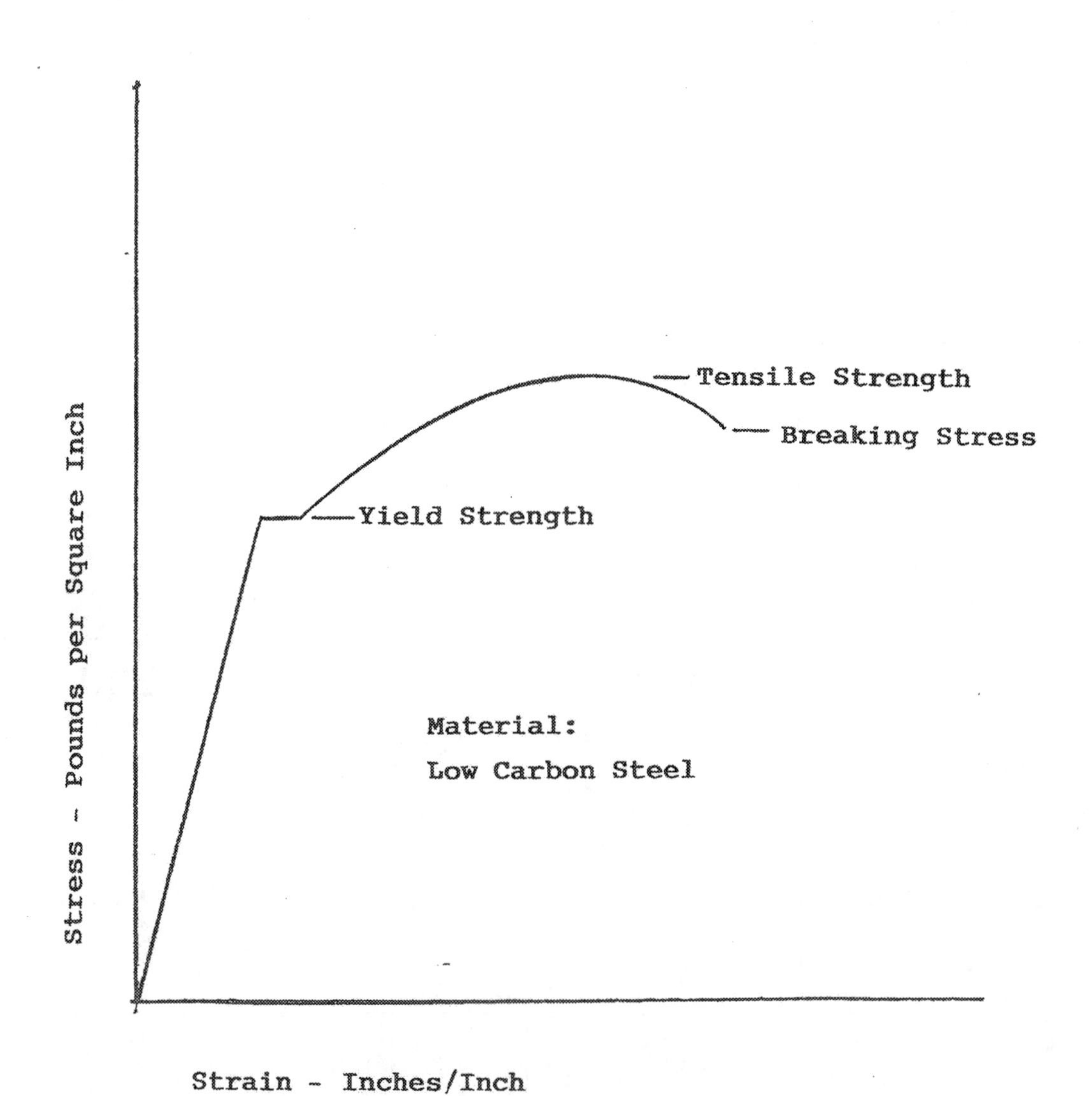

ENGINEERING DRAWING

Geometric Dimensioning and Tolerancing describes a dimensioning system that is in use for preparing engineering drawings. It does not replace conventional dimensioning and tolerancing but supplements it by more precisely controlling requirements. It increases the tolerance zone for machining some features such as drilled holes and more precisely controls the fit between mating parts so that assemblies can be made without costly delays in production.

The sketch at the top of Figure 2 has a plate with a drilled hole. It is dimensioned using conventional dimensioning and tolerancing drawing practice. Assuming that the tolerance found in the drawing block for the 2.5 and 5.0 dimensions is plus or minus .25; the tolerance zone for the drilled hole is a square with sides equal to .5.

The lower sketch on Figure 2 is the same component with Geometric Dimensioning and Tolerancing. It contains the same information as the upper sketch along with a boxed in feature control frame that is applied to the hole. The crossed circle is the geometric characteristic symbol for position tolerance. The slashed circle indicates diameter. The .7 is the diameter of a circle that the center of the hole must fall within. It equals 1.4x.5 which is the distance from one corner of the .5x.5 square to the opposite corner established above using conventional tolerancing. The circled M is the material condition symbol meaning "maximum metal" (smallest hole). Other material condition symbols are L "least metal" or largest hole and R ("regardless of feature size"). A, B, and C are reference datums applied to plate boundaries with A being the primary, B being the secondary, and C being the tertiary reference surface. The locational tolerance for larger hole centers is greater than for smaller holes. This system of dimensioning allows more area for the location of the hole center and more complete definition of the hole location with reference to all three datums insuring the successful assembly of the plate to its mating components.

Figure 2

Engineering Drawing

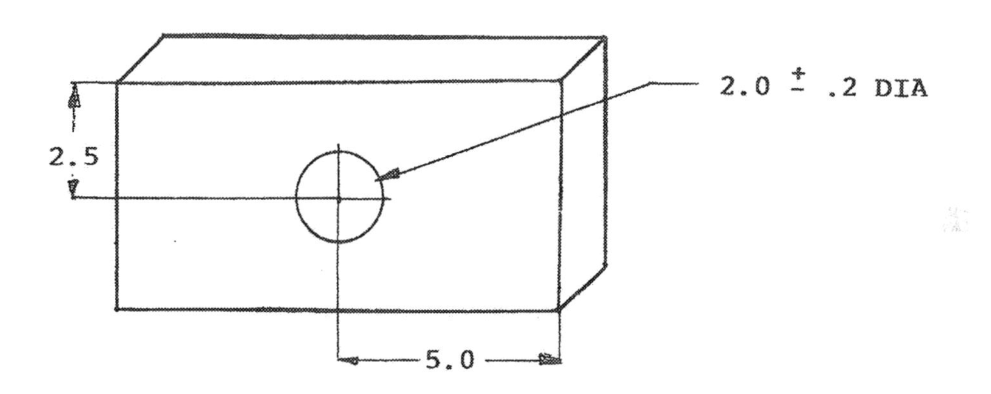

Traditional Drawing Practice

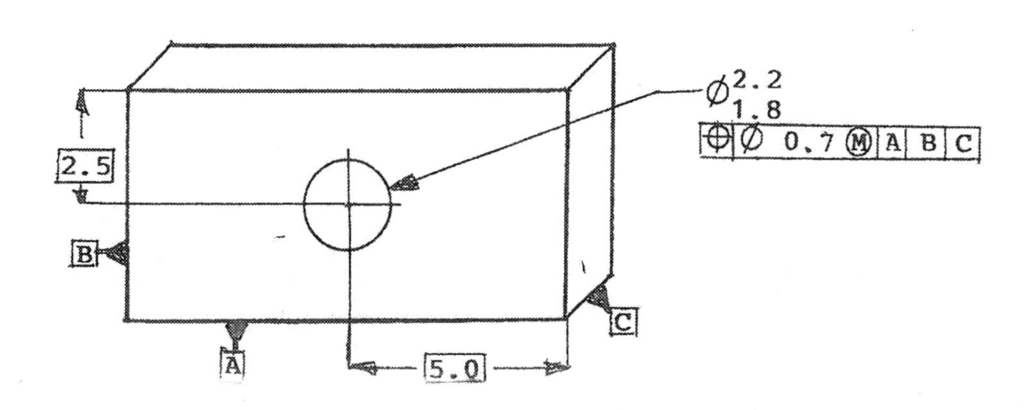

Geometric Dimensioning & Tolerancing

FASTENERS AND COUPLINGS

Screw Fasteners: There are two major standards for screw threads: Unified Inch Screw Threads and Metric Screw Threads. Unified Inch (UN) Screw Threads were adopted by Canada, the United Kingdom, and the United States. They come in various series and tolerance classifications. Among them are coarse, fine and extra fine series and 1, 2 and 3 tolerance classes. Coarse series fasteners are for general use while fine series are for automotive, aircraft or where maximum strength is required. Number 1 is the liberal tolerance for easy assembly; number 2 is for general assembly, and number 3 is for accurate assembly. Nominal sizes range from 0 to 6 with 0 having a major diameter of 0.0600 inches and 6 having a major diameter of 6.0000 inches. The Metric screw threads are under the sponsorship of the International Organization for Standardization (ISO). The basic profile of the thread is the same as Unified Screw Threads. They have various tolerance grades and come in coarse and fine pitch sizes ranging from 0.25 mm to 45 mm. Bolts are also graded according to strength requirements. The Society of Automotive Engineers (SAE) numbers the grades 1 through 8 (metric 4.6 through 12.9). Number 1 grade specifies low or medium carbon steel with a tensile strength of 60,000 psi while the highest number 8 grade specifies alloy steel and has a tensile strength of 180,000 psi.

Couplings: A coupling is a mechanical component that is used to connect the ends of two shafts together so that power can be transmitted through the juncture. Rigid couplings are used on shafts that are in perfect alignment. At the top of Figure 3 is a sketch of a rigid coupling with two flanged members. When they are bolted together, tapered keys are wedged between the coupling and the shafts locking the two shafts together. Flanged couplings can be assembled to the shafts with loose fitting keys if the flanges are a press fit on the shaft. Flanged coupling are used in wind turbine drive trains. For shafts that are not closely aligned, a flexible coupling is used. The middle sketch of Figure 3 is of a double slider flexible coupling. The two end pieces are attached to their shafts while the center piece is free to move to compensate for shaft misalignment and offset. For shafts with a high amount of misalignment, a flexible coupling called a universal joint is used. The lower sketch of Figure 3 is of a universal joint with the two end pieces connected together by a needle bearing mounted center cross piece. This type of coupling is frequently used in pairs to maintain a constant velocity output.

Figure 3

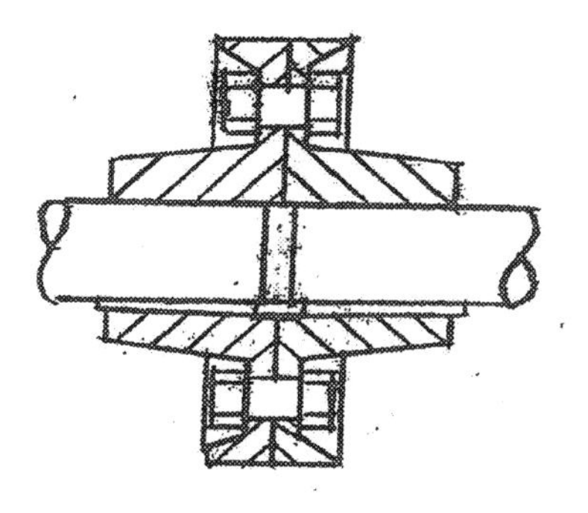

Flanged Coupling

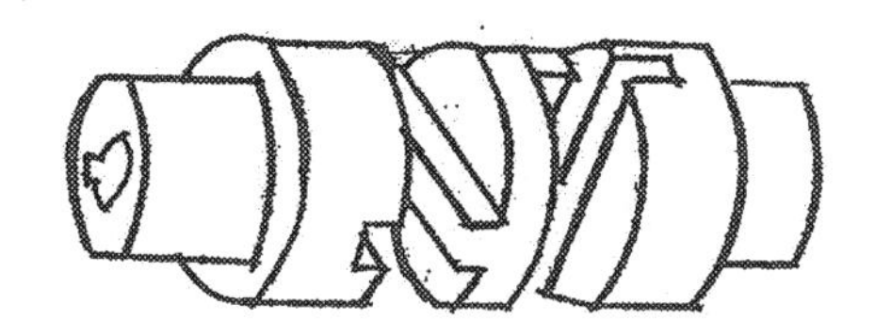

Double Slider Coupling

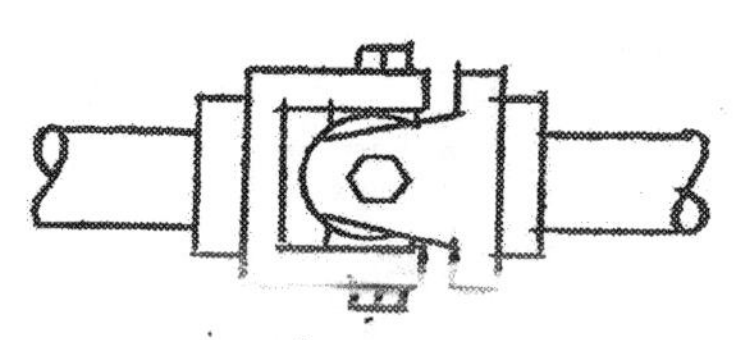

Universal Joint

BELTS AND PULLEYS

Belts and pulleys are used when the distance between shafts is too great to make it practical to use gears. There are basically two kinds of belts: flat belts and V belts. Flat belts, as the name suggests, are wider and thinner than V belts which have a V shaped section. Both kinds are made of fabric which is impregnated with rubber. Belts are a quiet and efficient means of transmitting power and are able to absorb vibration. A cross section of a flat belt is shown at the top of Figure 4 while a section of a V belt and pulley is shown at the bottom. Flat belts have a crown shaped surface which acts to keep the belt centered. V belt pulleys have an internal angle slightly less than the belt. This produces a wedging action of the belt into the pulley aiding traction. Generally speaking, flat belts are as efficient and can run as fast as gears, but deliver only a fraction of the power. V belts are less efficient but can run only about 60% of the speed of gears and deliver a fraction of the power.

V belts are used extensively in the front end of automobile engines to power various vehicle accessories. At the top of Figure 5 is a diagram of a typical vehicle engine accessory drive system. It can be seen that the engine crankshaft drive pulley has one belt that drives the fan pulley and the power steering pulley and a second that drives the fan pulley and the alternator pulley. The fan pulley has a third belt that drives the air-conditioner pulley.

At the bottom of Figure 5 is a diagram of a typical engine accessory drive system using a single Poly V belt. It can be seen that it is wider and thinner than a standard V belt and, as the name suggests, has a number of smaller V belt gripping surfaces instead of the one large surface characteristic of standard V belts. Poly V belts have a distinct advantage over standard V belts in automotive accessory drive systems. They are stronger, more flexible, use smaller diameter pulleys, and last longer than standard V belts. Also, the back of poly V belts acts like a flat belt and can also be used to drive pulleys. Because of these advantages, the design package using poly V belts is smaller and simpler than using standard V belts. It is common practice to use a belt tensioner with poly V belts in the front end of automotive vehicles but the advantages far outweigh any disadvantage this may have.

Figure 4

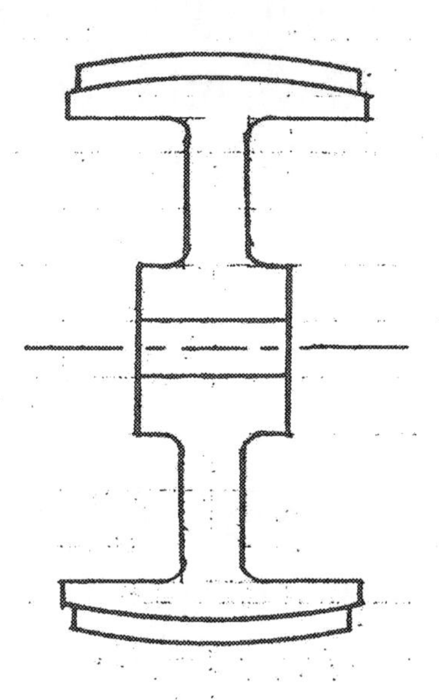

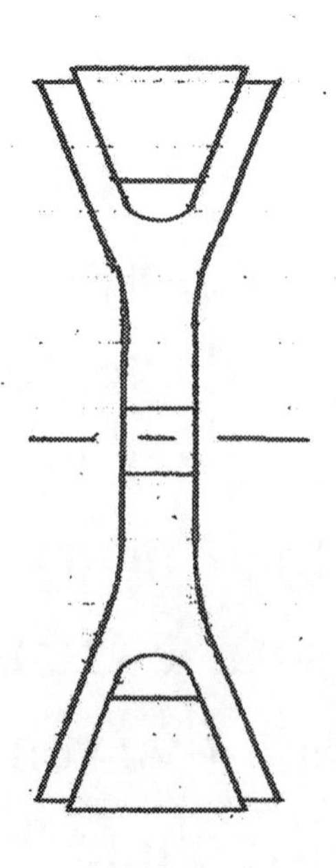

Figure 5

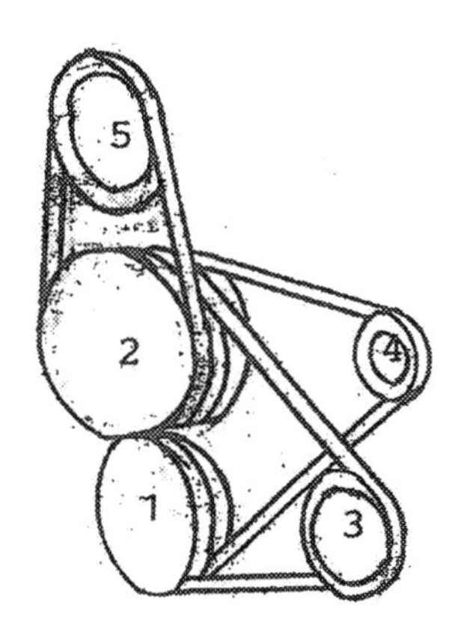

Three V Belt System

#1 is the crankshaft drive pulley.
#2 is the fan pulley.
#3 is the power steering pulley.
#4 is the alternator pulley.
#5 is the air conditioner pulley.

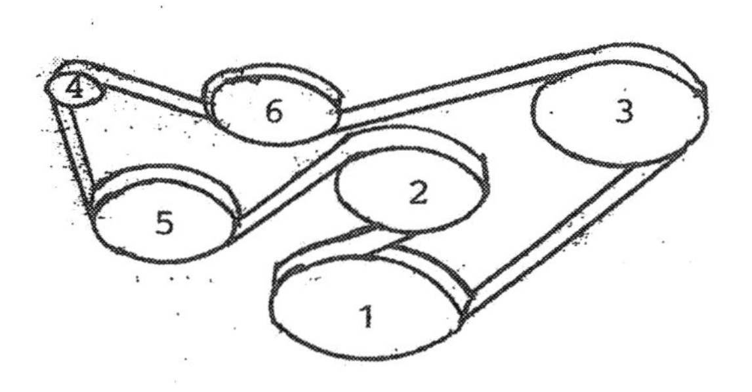

Single Poly V Belt System

#1 is the crankshaft drive pulley.
#2 is the fan pulley.
#3 is the power steering pulley.
#4 is the alternator pulley.
#5 is the air conditioner pulley.
#6 is the belt tensioner pulley.

BEAM FORMULA

A beam is a structural member that can be supported at one end (cantilever) or both ends (simple). Many formulas are available in engineering text books that are used to calculate stresses and deflections of beams under various types of loading. Beam formulae can be used as a good design tool for various engineering applications. One such formula was used to calculate the moment load on the shaft supporting a gear as shown on Figure 6. The moment load is needed to determine if the shaft is adequately sized for the horsepower output of the gear. The equation follows:

$$M=W(2c+b)[4al+b(2c+b)/8l^2$$

M is the moment load on the gearshaft in inch-lbs. It acts on the center of the gearshaft. W is the load on the gearshaft in pounds which is produced by the separating force exerted by the spur gear (shown) driving against its mating gear (not shown). c is the distance from the right side of the gear to the right bearing in inches. b is the width of the gear. a is the distance from the left side of the gear to the left bearing. l is the distance between the two bearings.

The above formulae can be rearranged as follows to calculate the maximum load that the shaft can support as:

$$W=8l^2fS/(2c+b)[al+b(2c+b)]$$

f is the shaft extreme fiber bending stress in pounds per square inch. S is the shaft section modulus in $inches^3$.

Another beam formula can be used to calculate the misalignment of the two shaft support bearings under load. Excessive misalignment can cause premature bearing failure. It is as follows:

$$Q=Wl^2/24EI$$

Q is the bearing misalignment in inches per inch. W is the load as above in pounds. l is the shaft length between the bearings as above. E is the shaft material modulus of elasticity. I is the shaft moment of inertia.

Figure 6

Beam Formula
Applied to Gear Mounting

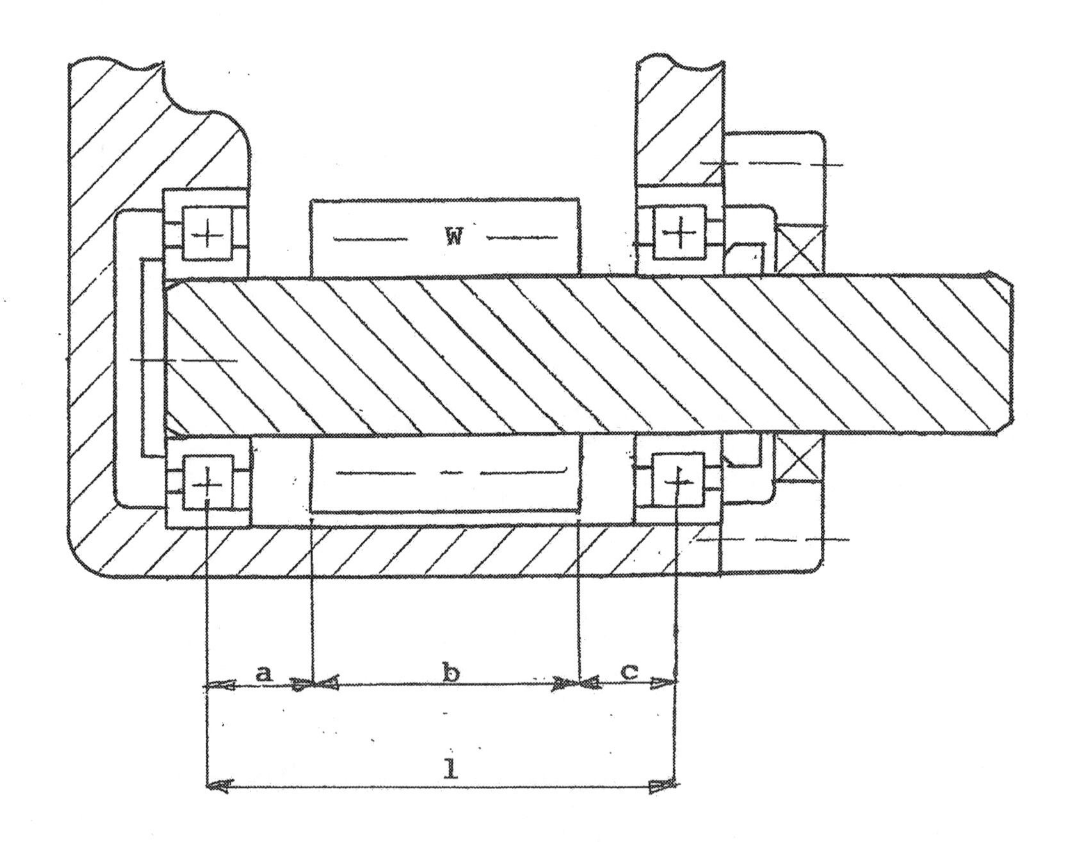

Spur Gear Drive Pinion

SHAFTING DESIGN

An article courtesy of Mechanical Engineering magazine Volume 49/No.5, May, 1927 474-476; copyright Mechanical Engineering magazine (The American Society of Mechanical Engineers) covers design formulas for cases most frequently met in design of power transmission shafting. The formulas in the report are general and apply to all cases of shaft design in which strength and not deformation is the prime factor. The following formulae transposed down to one line, and without the modifying factors referenced in the above mentioned article, apply to solid shafts subjected to torsion, bending, and axial loading: (In actual practice the modifying factors must be used according to the above-mentioned technical publication.)

$$D=\{16/\pi p_t[M+(FD/8)]^2+T^2\}^{1/3}$$

D is the outside diameter of the shaft in inches. p_t is the maximum shaft shear stress in pounds per square inch. M is the maximum bending moment in inch-pounds. F is the axial load in pounds. T is the torque in inch-pounds.

The above formulae can be used to calculate the diameter of the shaft shown in Figure 6. For a spur gear there will be imposed shear, bending, and torque loading. For a helical gear there will also be an axial load. The following formula is used where the shaft is subject to torsion and bending only:

$$D=[16/\pi p_t(M^2+T^2)]^{1/3}$$

The following formula is used for shafts in pure torsion only:

$$D=(321{,}000P/S_tN)^{1/3}$$

P is the maximum transmitted power by the shaft in horsepower. S_t is the maximum permissible torsional shear stress in pounds per square inch. N is the shaft speed in revolutions per minute. The following formula is for shafts in bending only:

$$D=(32M/\pi S)^{1/3}$$

S is the maximum bending stress in inch-pounds

ANTI-FRICTION BEARINGS

Anti-friction bearings utilize balls or rollers that rotate between an inner and outer ring to reduce friction in rotary equipment as opposed to sleeve bearings that incorporate a stationary tubular-shaped component. (See Figures 7, 8, & 9.) Anti-friction bearings operate in the high 90% efficiency range and are used in a wide variety of rotary equipment. In most applications there are two bearings (ball or roller) supporting a rotating shaft. Loads, or forces, are imposed on the bearings by the equipment that is driving or being driven by the shaft. The sketch at the top of Figure 10 illustrates that radial loads act perpendicular to the shaft and thrust loads act parallel to the shaft. In some instances there are two radial loads acting 90 degrees apart. The Pythagorean Theorem is used to calculate the resultant radial load. The radial load can sometimes be straddle mounted between two bearings as shown on the third sketch of Figure 10. Simple beam formulae will show that the bearing nearest the load supports the greater portion of the load.

The fourth sketch illustrates the load being overhung. Beam calculations show that the bearing closest to the load supports a force that is actually greater than the load itself. Bearing loads are inserted into the following equation that, along with speed of rotation, is used to predict service life:

$$L_{10}=3000(C/P)^{10/3}(500/S)$$

L_{10} is the life in hours that 90% of the bearings are expected to endure. C is the capacity of the bearing in pounds and is found in industry catalogs and is the number of pounds that the bearing can support for 3000 hours of operation at 500 rpm. P is the equivalent radial load which takes into account both radial and thrust loads imposed by the application and is found in industry catalogs. S is the application speed in revolutions per minute.

Researchers have found that, in some instances, a film of oil forms between the rolling elements and the rings preventing full metal-to-metal contact reducing friction and prolonging service life over calculated values. Oil viscosity, bearing operating speed, bearing load, and surface finish are prime factors in the development of the oil film. Charts are available that assist in predicting the thickness of the oil film and its effect on bearing life under a wide range of operating conditions. Depending on the existence and thickness of the oil film bearing life can be less than or more than calculated values. This field of endeavor is called Elastohydrodynamic lubrication.

Figure 7

Ball Bearing Terminology

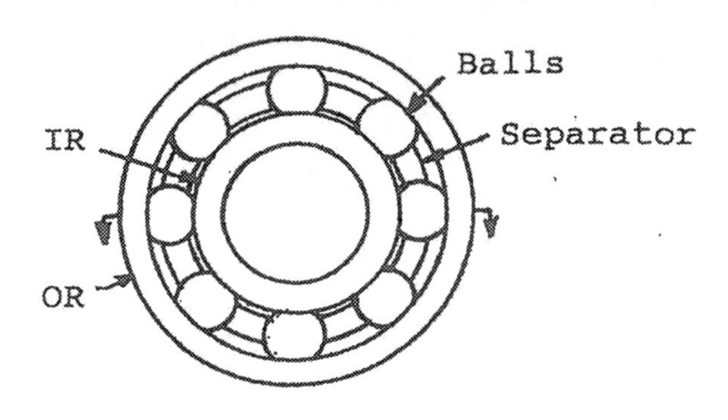

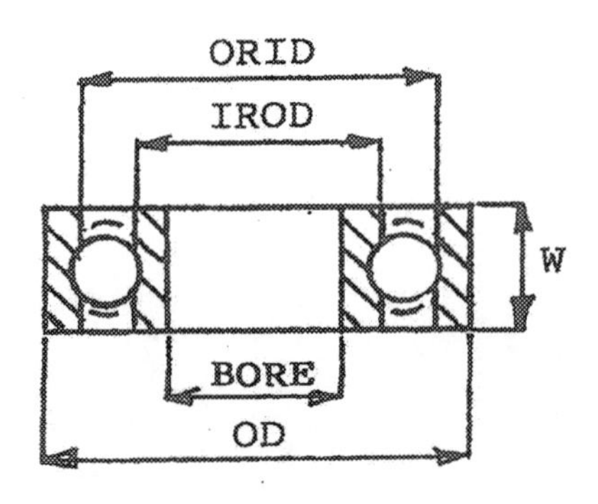

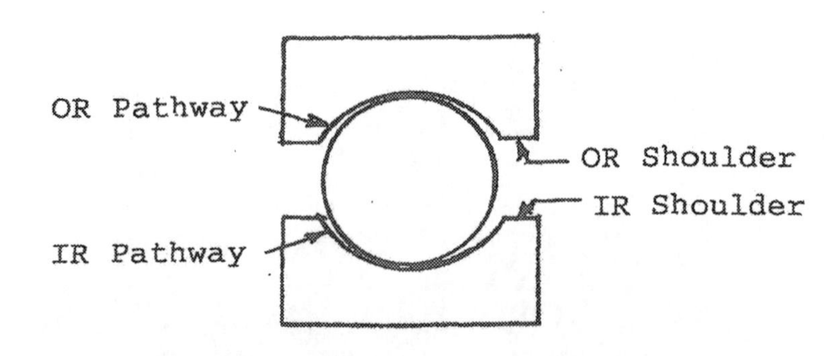

(Exaggerated View)

Figure 8

Cylindrical Roller Bearing

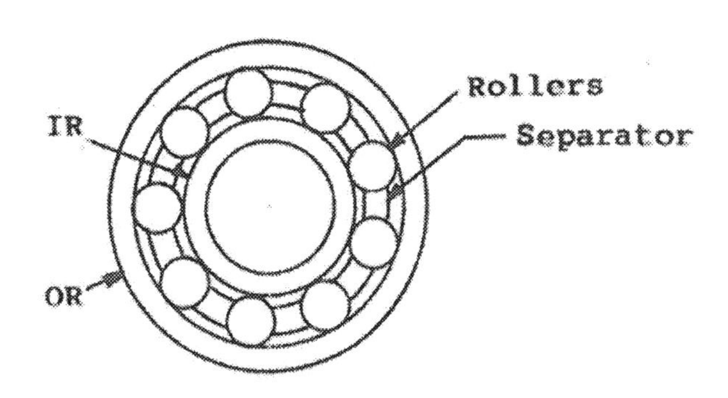

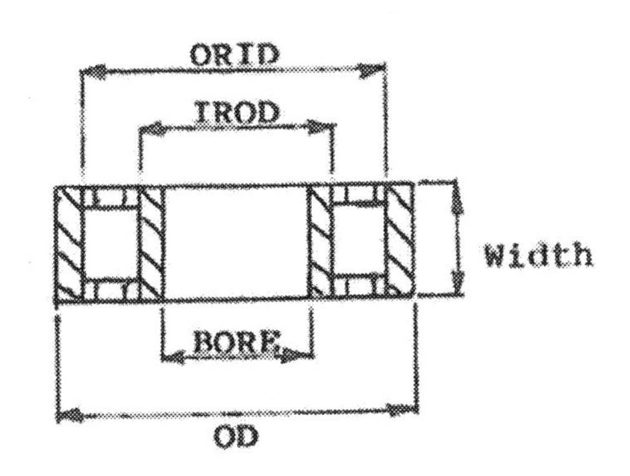

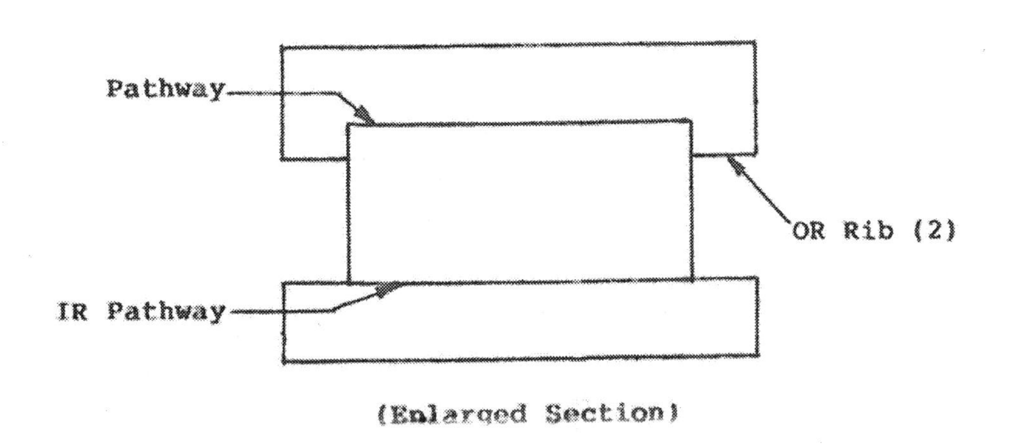

(Enlarged Section)

Figure 9

Tapered Roller Bearing

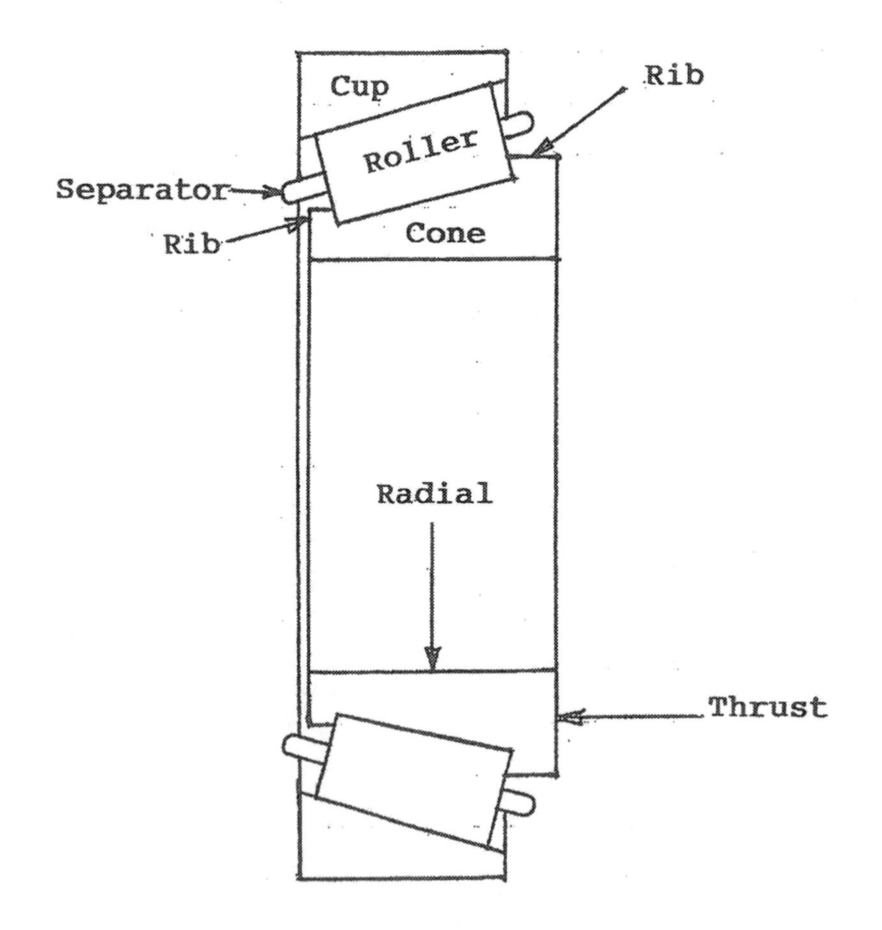

Figure 10

Bearing Loads

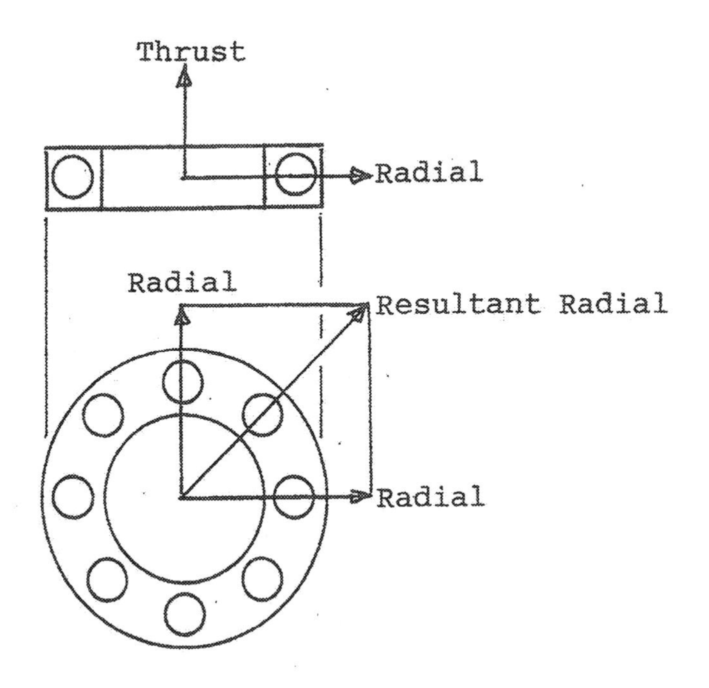

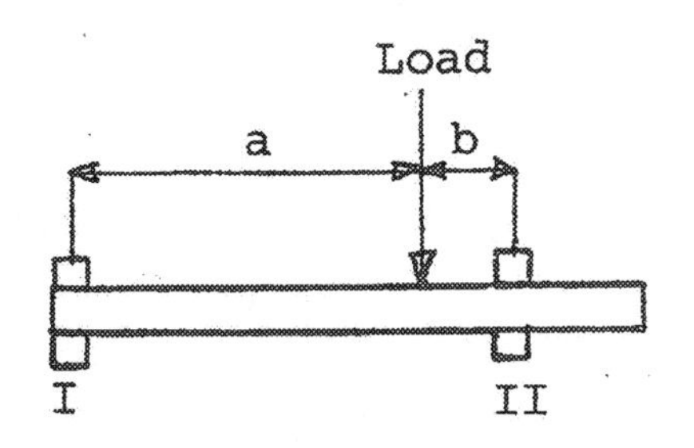

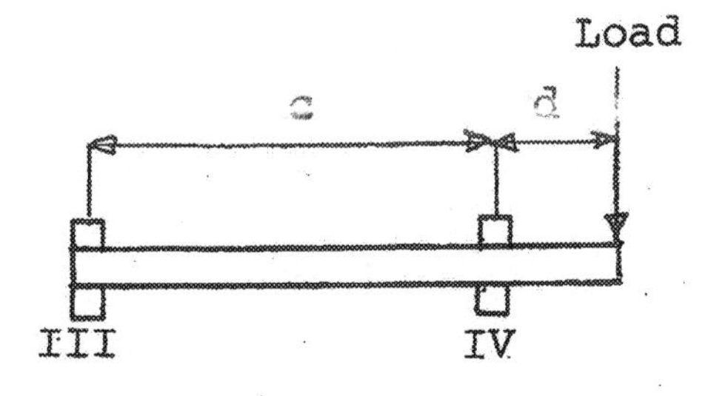

Bearing Lubrication: Highly refined mineral oils are among the best lubricants for rolling contact bearings. Synthetics have been developed that are good but some do not form elastohydrodynamic films as well as mineral oils. The best overall lubricating system is oil jet combined with a recirculating system. This method directs a pressurized stream if oil to the bearing load zone. The oil is then drained back to a sump where it is filtered, cooled and returned. The oil bath method is commonly used in gear boxes. The housing is filled with oil until it touches the lowest rotating member and splashed throughout the gearbox. See Figure 11 to determine the correct oil viscosity for any application. Draw a line representing the bearing bore in millimeters times the speed in rpm to the reference line and then down to the bearing operating temperature. Read oil viscosity at 100°F at the intersection. An effective means of lubricating bearings is by using grease. A carefully measured amount of grease is evenly distributed throughout a ball bearing and contained by seals or shields. This configuration is exceptionally advantageous to the machine designer because it can run for the life of the bearing without exceeding the envelope of a bearing without seals or requiring further maintenance.

Bearing Application: Figure 12 has a flexible shaft handpiece and drive unit, both using ball bearings. The handpiece must be small enough to fit inside a clasped hand and support loads from a variety of different tools that are attached to the left end. The handpiece right ball bearing is clamped both on the shaft and in the housing to locate the shaft and support thrust loads. The smaller left bearing is free to float in the housing to allow for manufacturing tolerances and shaft thermal expansion and is used only for radial support. The clamped spacer between the two bearings adds rigidity to the relativity small shaft. Three of the four drive unit ball bearings have seals at one end to retain lubricant and protect against contaminant entry.

Figure 13 has a machine tool workpiece center support with a double row ball bearing on the left and two single row ball bearings on the right. The double row ball bearing has ball/ring contact lines internally (toward the shaft centerline) diverging which serve to support radial and bi-directional thrust loads. The two single row ball bearings have contact lines in tandem (parallel) to support high one direction thrust loads. Both the left and right bearing inner rings are clamped against the spacer while the outer rings are free to float in the housing account for machining tolerances and shaft thermal expansion.

Figure 11

Bearing Lubrication

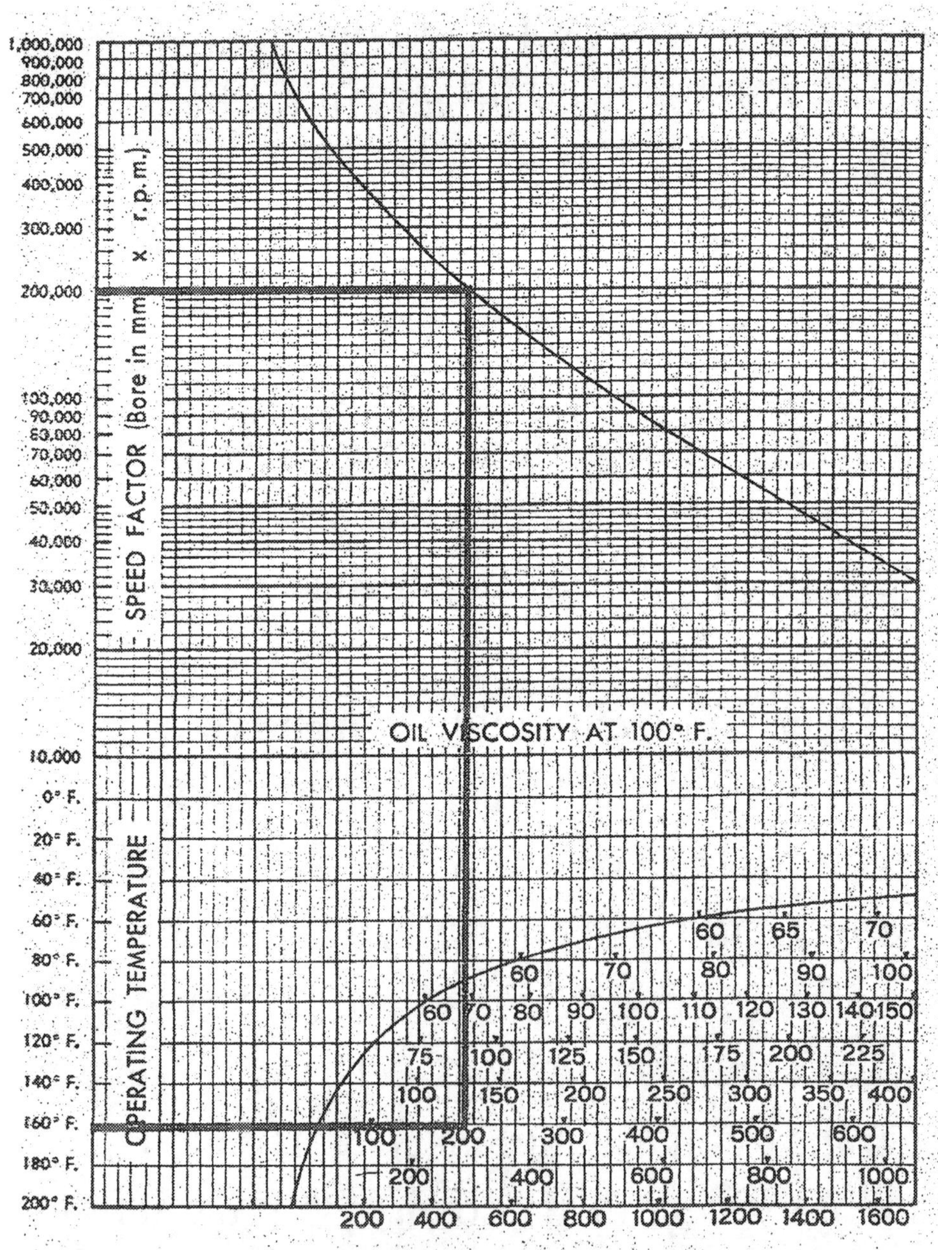

1) Multiply bearing bore in mm by rpm.
2) Draw horizontal line to reference line.
3) Draw vertical line to bearing temperature.
4) Read oil viscosity at intersection.

Figure 12

Radial Ball Bearing Application

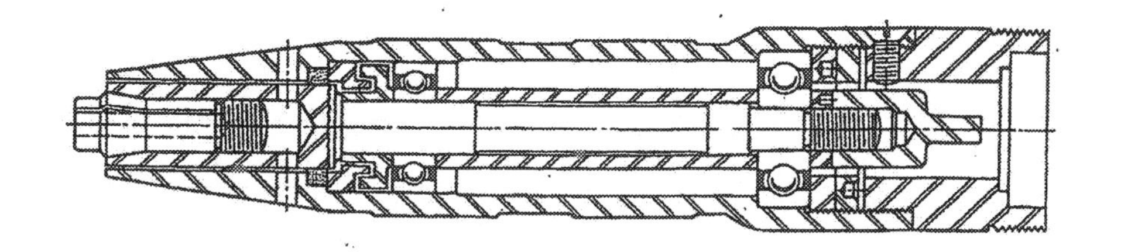

Flexible Shaft Handpiece

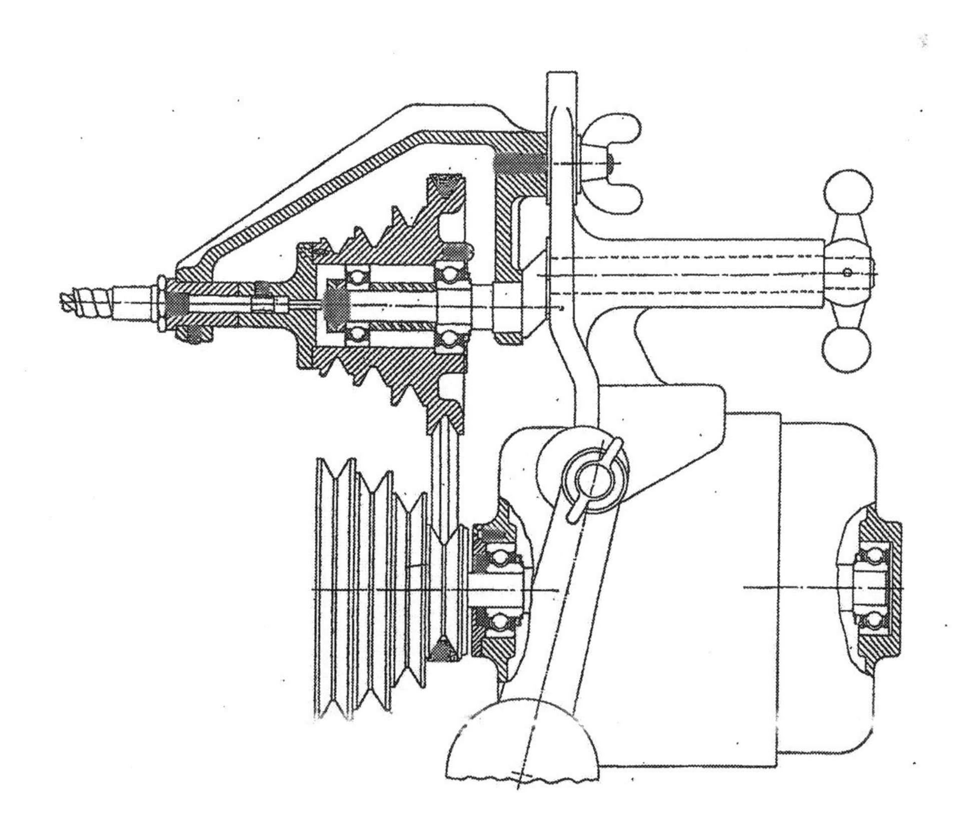

Flexible Shaft Drive Unit

Figure 13

Machine Tool Workpiece Center Support

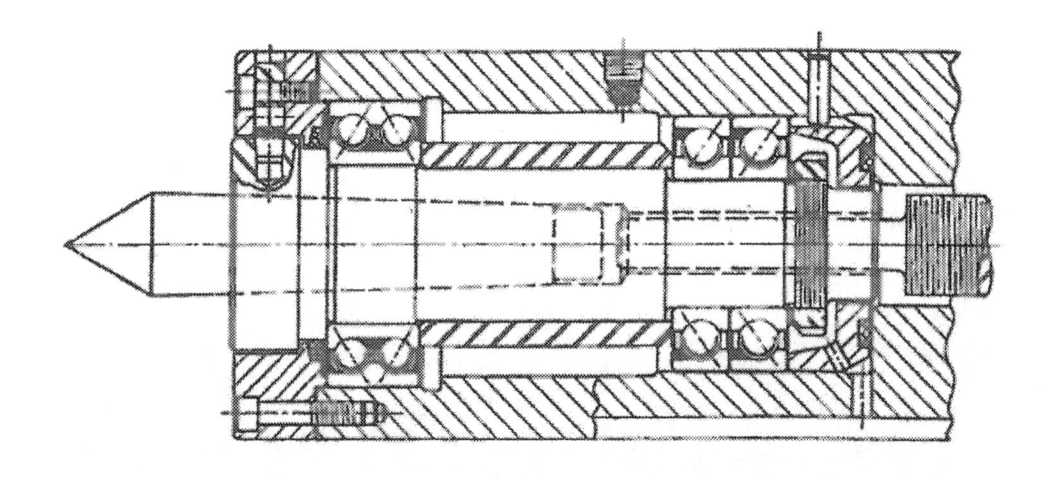

Double Row and Angular Contact Ball Bearings

Cylindrical roller bearings support a spur gearset shown on Figure 14. The roller bearings with double-ribbed outer rings and single-ribbed inner rings are designed to provide good bearing gear axial positioning and excellent radial support. Bearing inner and outer rings can both be assembled with press fits on their respective mounting surfaces easing assembly. The right side of the drawing shows an alternate method of bearing mounting below the centerline using end caps. This method allows thru-boring the housing for better gear and bearing alignment. It also allows removal of the pinion gear without separating the main housing.

The center section of an automotive drive axle is shown on Figure 15. Power is delivered from the shaft on the right to the top and bottom output shafts on the left. Nut preloaded tapered roller bearings support the input shaft while shim preloaded tapered roller bearings support the output shafts. The center section incorporates two sets of differential gears that rotate on sleeve bearings. This assembly allows power to be delivered to the two output shafts even though one may be rotating faster than the other such as when a vehicle is rounding a corner.

Figure 14

Spur Gearset With Roller Bearings

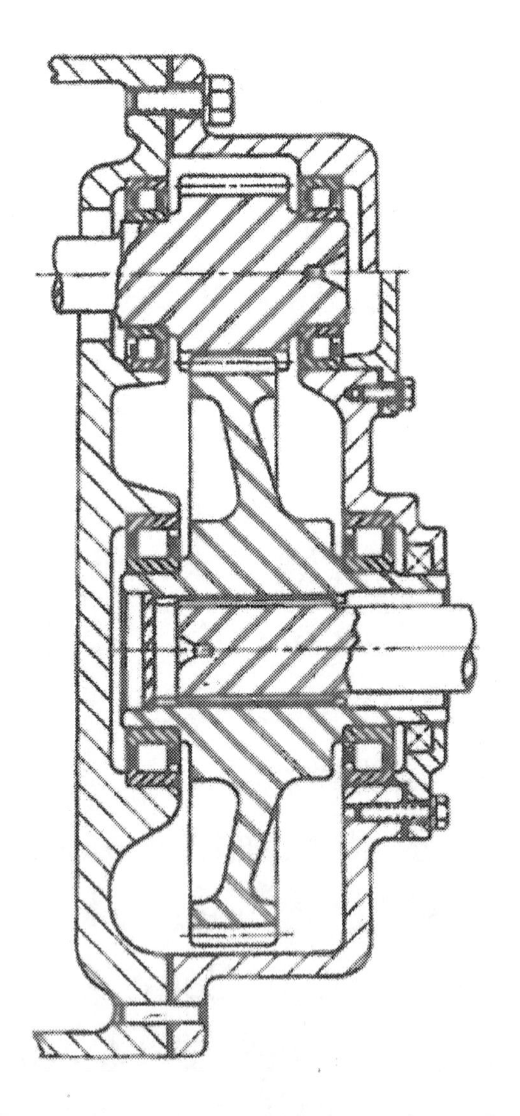

Optional Bearing Mounting Shown on Right

Figure 15

Automotive Drive Axle

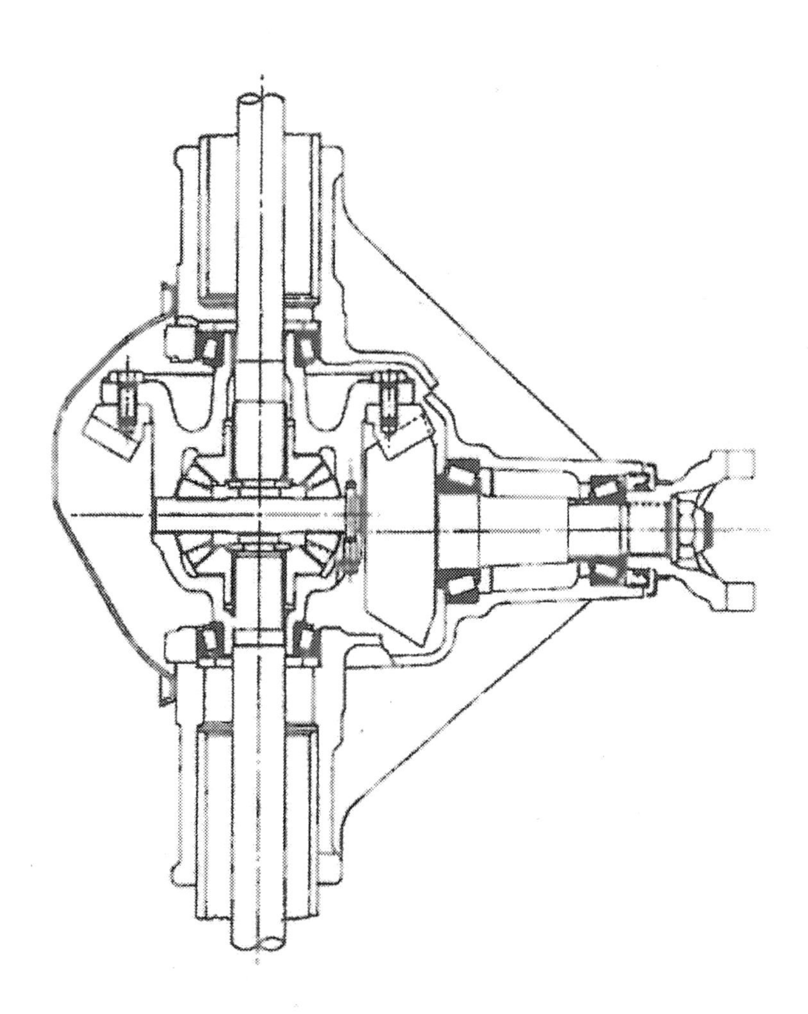

Axle Center Section

INVOLUTE GEARS

Gears are one of the most important components in the field of “Mechanical Power Transmission”. The gear tooth has been so successfully perfected that, when two gears mesh, almost perfect rolling takes place. Most gears operate in the high 90% efficiency range similar to anti-friction bearings. By changing the diameter of one gear with respect to another, they can be designed to regulate rpm and torque. A gear that is being driven by a smaller gear (pinion) 3/4 its own size will rotate at 3/4 the speed of the smaller speed and deliver 4/3 the torque as seen on Figure 16 between the driver and idler gear. The idler gear, having the same number of teeth as the drive gear, serves only to change the direction of rotation between the drive and driven gears. Precaution has to be taken when using idler gears because the teeth undergo reverse bending which shortens their lives compared to the drive and driven gears where only single direction bending takes place. The advantageous use of gears is exhibited in the transmission of an automobile where they are used to power the vehicle in a very efficient manner.

Gear Tooth Terminology: Figure 17 has a sketch with basic gear tooth terminology. Following are additional terms associated with gearing:

- Pinion is the smaller of two gears in mesh. The larger is called the gear regardless of which one is doing the driving.
- Ratio is the number of teeth on the gear divided by the number of teeth on the pinion.
- Pitch Diameter is the basic diameter of the pinion and the gear which when divided by each other equals the ratio.
- Diametral Pitch is a measure of tooth size. It equals the number of teeth on a gear divided by the pitch diameter in inches. Diametral pitch can range from 1/2 to 200 with the smaller number indicating a larger tooth.
- Module is a measure of tooth size in the metric system. It equals the pitch diameter in millimeters divided by the number of teeth on a gear. Module equals 24.400 divided by the diametral pitch. Module

can range from 0.2 to 50 with the smaller number indicating a smaller tooth.

- Pitch Circle is the circumference at the pitch diameter.
- Circular Pitch is the distance along the pitch circle from a point on one gear tooth to a similar point on an adjacent gear tooth.
- Addendum of a tooth is its radial height above the pitch circle. The addendum of a standard proportion tooth equals 1.000 divided by the diametral pitch. The addendum of a pinion and mating gear are equal

Figure 16

Gears Delivering
Motion and Force

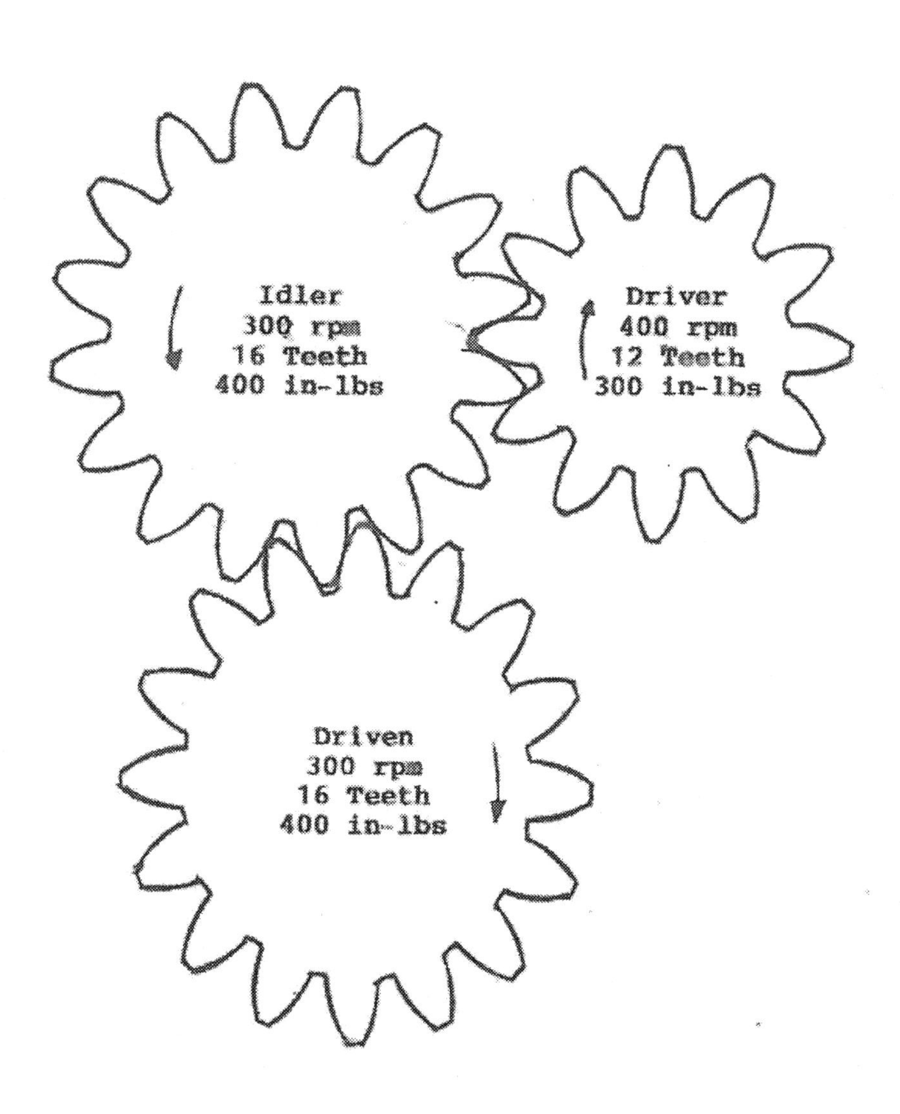

Figure 17

Gear Tooth Terminology

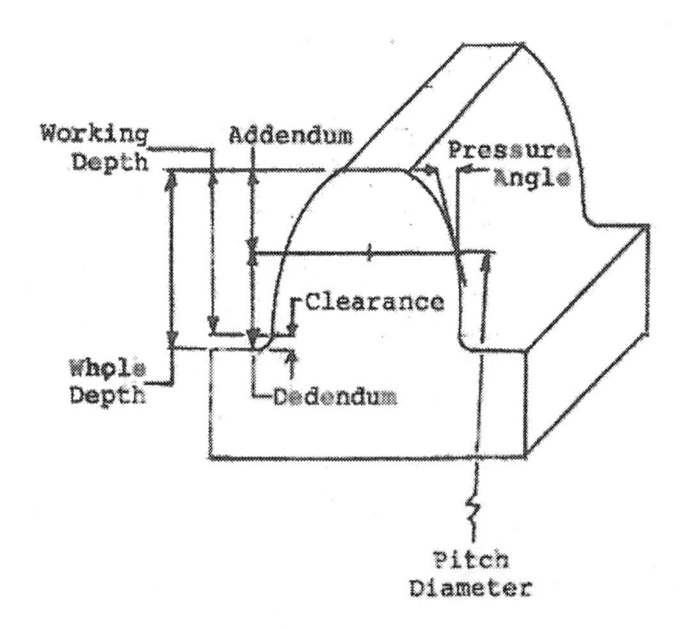

Dimensions For a 2 Diametral Pitch Tooth:

Addendum = 1.00/2 = 0.500 inches
Dedendum = 1.25/2 = 0.625 inches
Clearance = 0.25/2 = 0.125 inches
Whole Depth = 2.25/2 = 1.125 inches
Working Depth = 2.00/2 = 1.000 inches

except for the "Long Addendum" design where the pinion addendum is increased while the gear addendum is decreased by the same amount.

- Dedendum of a tooth is its radial height below the pitch circle. The dedendum of a standard proportion tooth equals 1.250 divided by the diametral pitch. The dedendum of a mating pinion and gear are equal except in the long addendum design where the pinion dedendum is decreased while the gear dedendum is increased by the same amount.
- Whole Depth or total depth of a gear tooth equals the addendum plus the dedendum. The whole depth equals 2.250 divided by the diametral pitch.
- Working Depth of a tooth equals the whole depth minus the height of the radius at the base of the tooth. The working depth equals 2.000 divided by the diametral pitch.
- Clearance equals the whole depth minus the working depth. The clearance equals the height of the radius at the base of the tooth.
- Pressure Angle is the slope of the tooth at the pitch circle.

<u>Types of Gears:</u> Four basic types of gears commonly used are spur, helical, bevel, and spiral bevel. (See Figure 18.)

A spur gear has teeth that are uniformly spaced around the outer surface. The teeth are aligned in a direction that is parallel to the gear axis. The shape of the contacting faces of gear teeth is in the form of an involute curve which is the same as that generated by a string that is unwound from a cylinder. Spur gears are designed to mesh with another spur gear on a parallel shaft. They impose radial loads only on the shaft. They are the most economical and commonly used type of gear used today.

Helical gears are like spur gears except that, instead of the teeth being parallel to the gear axis, they are aligned at an angle across the outer surface of the gear. The angle is called the helix angle and normally ranges from 10° to 30$^{o.}$ Helical gears run smoother and quieter than spur gears and can support a higher load but at a lower efficiency. Helical gears impose both radial and thrust loads on the shaft.

Bevel gears are used to transmit speed and torque between two shafts that are not parallel to each other such as 90°. They operate in the high 90% efficiency range. Spiral bevel gears have angled teeth that can be compared to bevel gears what helical gears are to spur gears.

Figure 18
Types of Gears

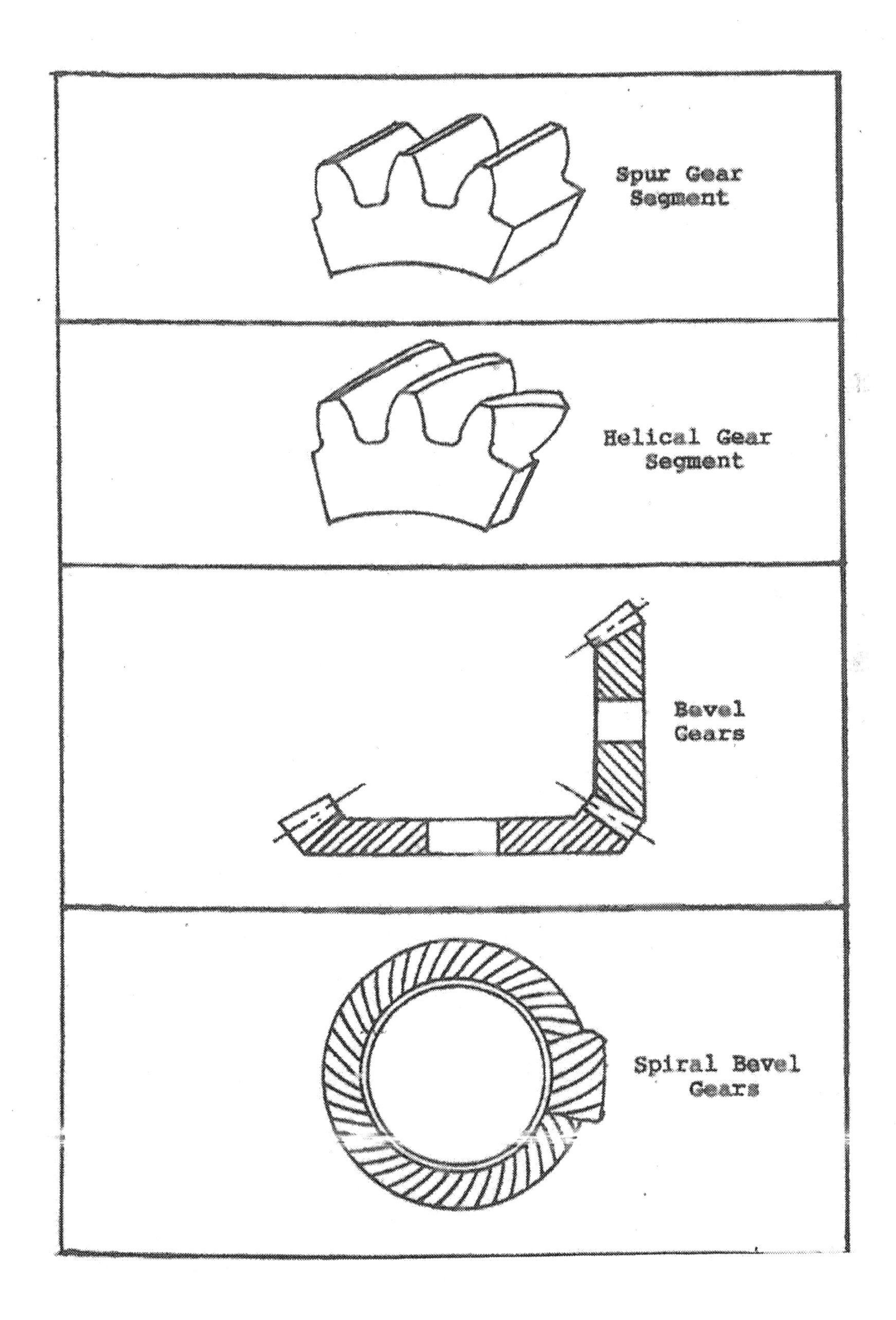

Gear Tooth Bending: Wilford Lewis calculated gear tooth bending strength in 1892. He was the first to treat the gear tooth as a cantilever beam. He conceived the idea of inscribing the largest parabola that would fit inside a gear tooth. Using the appropriate cantilever beam equation, he then calculated the stress which is constant all along the contour of the parabola. The weakest part of the tooth is where the parabola is tangent to the surface of the gear tooth. This point occurs near the base of the tooth where the involute curve meets the fillet radius. (See Figure 19.) The magnitude and location of the maximum bending stress on the tooth are now known.

The American Gear Manufacturers Association (AGMA) base their gear tooth bending stress power rating formula on the work of Lewis in Standard ANSI/AGMA 2001-D04. The formula transposed down to one line and without the modifying factors follows: (In actual practice, all the modifying factors must be used to insure accurate results.)

$$P_{at}=(\pi n_p dFJs_{at})/(396{,}000P_d)$$

P_{at} is the tooth bending strength allowable transmitted horsepower for 10 million cycles of operation at 99%

reliability. π (pi) is a constant. n_p is the pinion speed. d is the pinion operating pitch diameter in inches. F is the gear face width. J is the geometry factor for bending strength. s_{at} is the allowable bending strength number. P_d teeth diametral pitch which is a measure of tooth size.

Gear Tooth Pitting: The stress on the contact area of gear teeth is based on the work of German physicist Heinrich Hertz. He determined the stress and shape of the contact pattern between various geometric figures. Of interest for gear work is the stress and contact pattern between two parallel cylinders which simulates the condition existing between two mating gear teeth. AGMA bases their gear pitting resistance power rating formula on the work of Hertz which is also found in Standard ANSI/ AGMA 2001-D04:

$$P_{ac}=(\pi n_p FI/396{,}000)(ds_{ac}/C_p)^2$$

P_{ac} is the pitting resistance allowable transmitted horsepower for ten million cycles of 99% reliability. I is dependent on load sharing and the radius of curvature of the two contacting surfaces and is the geometry factor for pitting. $\mathbf{s}_{ac}$ is the allowable contact stress. Cp is the elastic coefficient.

Figure 19

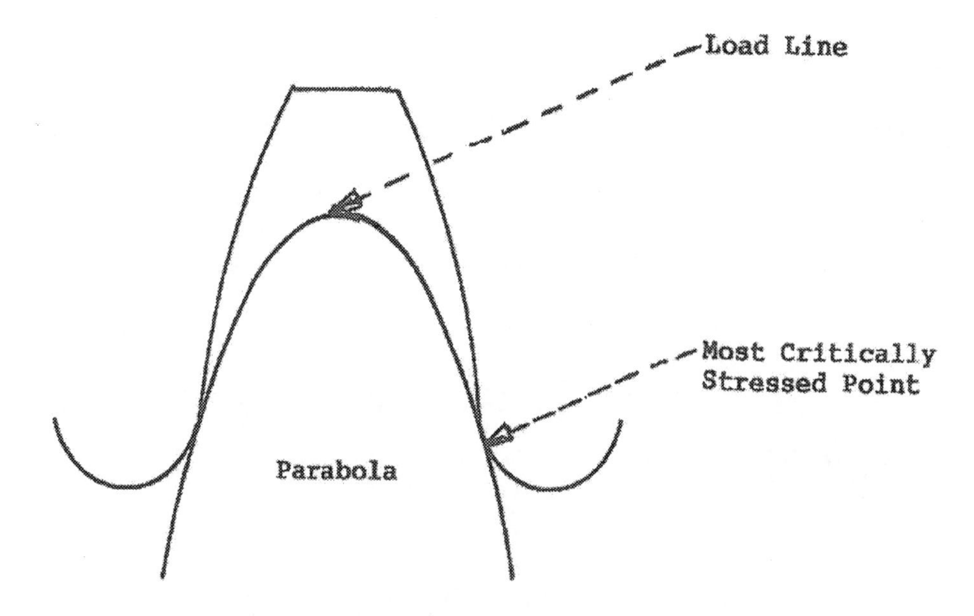

Lewis Method for Calculating
Gear Tooth Bending Stress

GEAR TRAINS

Gear trains are multiply sets of gears meshing together to deliver power and motion more effectively than can be accomplished by one set of gears. Figure 20 illustrates the various kinds of gears that can be used in a gear train. Gears 2 and 3 can be either spur or helical gears and are mounted on parallel shafts. Gears 4 and 5 are bevel gears that mount on shafts that are 90° apart. Gears 6 and 7 comprise a worm gear set and mount on shafts that are 90° apart but are non-intersecting. Worm gears have a high ratio and can be non-reversing.

Figure 21 has a simple gear train at the top and a compound gear train at the bottom. The simple gear train consists of four in-line gears in mesh. The compound gear train consists of the same four gears, except that two are located on the same shaft. The overall ratio of the simple gear train is the product of the three individual ratios:

$$n_2/n_5=(N_3/N_2)(N_4/N_3)(N_5/N_4)$$

n equals the rpm and N equals the number of teeth in the respective gears. When cancelling out like quantities, the equation reduces to the following:

$$n_2/n_5=(N_5N_2)$$

If gear 5 has 64 teeth and gear 2 has 16 teeth, the overall gear train ratio would be 4 to 1. If gear 2 is considered the driving member, the gear train in Figure 21 is a speed reducing gear train. If the rpm of gear 2 is 400, the rpm of gear 5 is 100. Gears 3 and 4 are called idler gears since they have no effect on the overall gear train ratio. The equation for the overall ratio of the compound gear train is the product of the two individual ratios:

$$n_2/n_5=(N_4/N_2)(N_5/N_3)$$

Assuming that the same four gears are used and that gear 3 has 32 teeth and gear 4 has 48 teeth, the overall ratio for compound gear train is:

$$n_2/n_5=(48/16)(64/32)=6$$

Figure 20

Kinds of Gears

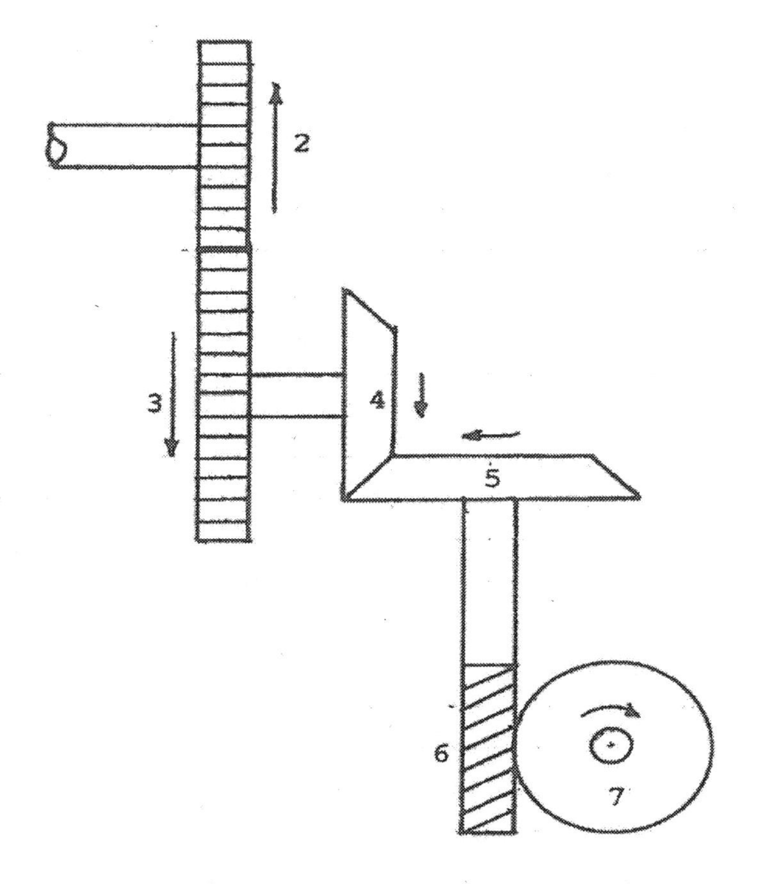

2/3 Spur or Helical Gears
4/5 Bevel or Spiral Bevel Gears
6/7 Worm Gear

Figure 21

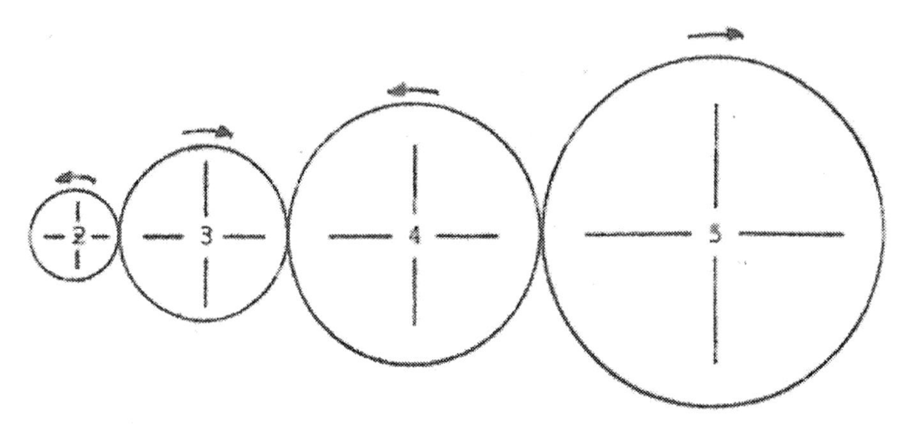

Simple Gear Train

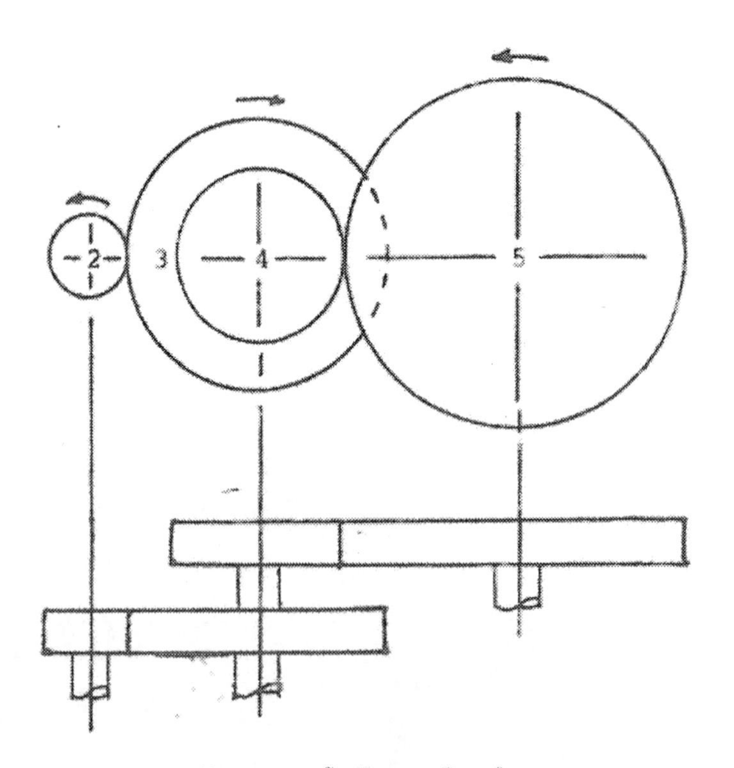

Compound Gear Train

INDUSTRIAL GEAR APPLICATION

Industrial gearboxes contain one or more pairs of gears inside a housing with input and output shafts. These gearboxes contain a gear lubricant, are sealed, and usually operate maintenance free. Connected to the input shaft is a high speed power source such as a motor and connected to the output shaft are devices which generally use lower rpms and higher torques to perform a particular task. The device may include an appliance, a machine tool, a conveyor, or an elevator.

Figure 22 has a worm gearbox at the top and bevel gearbox at the bottom. Worm gears have a threadlike shape or "worm" on the shaft which meshes with a gear. This configuration is used where a high reduction is required. The bevel gearbox at the bottom has input and output shafts 90° apart like the worm gearbox; however, bevel gearboxes axes intersect as opposed to worm gearboxes being offset.

Figure 23 has a double reduction gearbox at the top and a gearbox/motor combination unit at the bottom. The double reduction gearbox employs helical gears for higher torque delivery and quieter operation over spur gears. The gearbox/motor unit is an integral package that features a double reduction gearbox directly attached to a drive motor. Integral motor mounting features excellent alignment with the gearbox input shaft for smooth, vibration-free operation.

Figure 22

Worm Gearbox

Bevel Gearbox

Images courtesy of Emerson Power Transmission.

Figure 23

Double Reduction Gearbox

Gearbox With Motor

Images courtesy of Emerson Power Transmission.

AUTOMOTIVE POWER TRANSMISSION

Most today's automotive vehicles employ piston-type internal-combustion engines. Most passenger cars use spark-ignition gasoline fuel engines while larger vehicles use compression-ignition diesel fuel engines. The spark-ignition engine was invented by Nikolaus Otto of Cologne, Germany in 1876. Compression-ignition engines were invented by Rudolph Diesel also of Germany in 1893. At the top of Figure 24 is an Otto Cycle pressure vs. volume diagram characteristic of spark-ignition engines. From 0 to 1, with the intake valve open and the exhaust valve closed, the engine crankshaft retracts the piston sucking in a fuel/air mixture into the cylinder. From 1 to 2, with both valves closed, the piston advances compressing the fuel/air mixture to a higher pressure. From 2 to 3, the spark plug instantaneously ignites the mixture causing rapid heating at constant volume. From 3 to 4, with the exhaust valve open, rapid expansion of the gases retracts the piston causing work to be done on the crankshaft. From 4 to 1, the cylinder vents. From 1 to 0, the piston advances purging all residual exhaust gases from the cylinder after which time the cycle is repeated. Work multiplied by cycles per second equals power. Otto cycle engines have efficiencies well below 50% because of high heat and fan losses compared to electric motor driven vehicles where efficiencies are well above 50%. A major limitation of Otto cycle engines is pre-ignition when cylinder pressure is raised beyond the limit of the fuel causing a condition called "pinging".

At the bottom of Figure 24 is a plot of the Diesel Cycle. The Diesel cycle is similar to the Otto cycle except that the air-fuel mixture is ignited by compressing it to its ignition point rather than using an electric spark. It can be seen on the charts that diesel fuel compression ignition occurs at constant pressure compared to spark-ignition that occurs at constant volume. Diesel engines operate with higher cylinder pressures making them more powerful and efficient than comparable gasoline engines.

Figure 25 has a sketch of a torque converter. Torque converters are used in automotive vehicles where they are mounted between the engine and transmission. The main body is in the form of a torus (donut). On the left is the input shaft to which the engine flywheel is attached. The flywheel is connected to the right side of the torus which contains a circular row of blades called the impeller. The impeller, when rotated, and with the aid of centrifugal force, delivers oil across the narrow gap to the left side of the

Figure 24

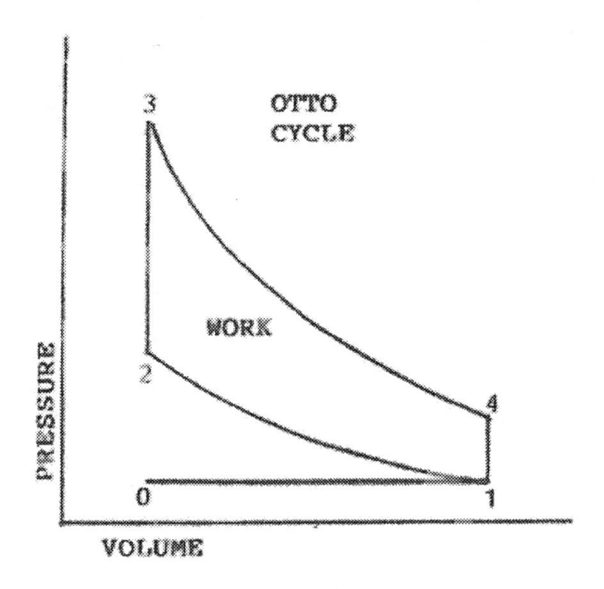
OTTO
CYCLE
WORK
PRESSURE
VOLUME
3
2
4
0
1

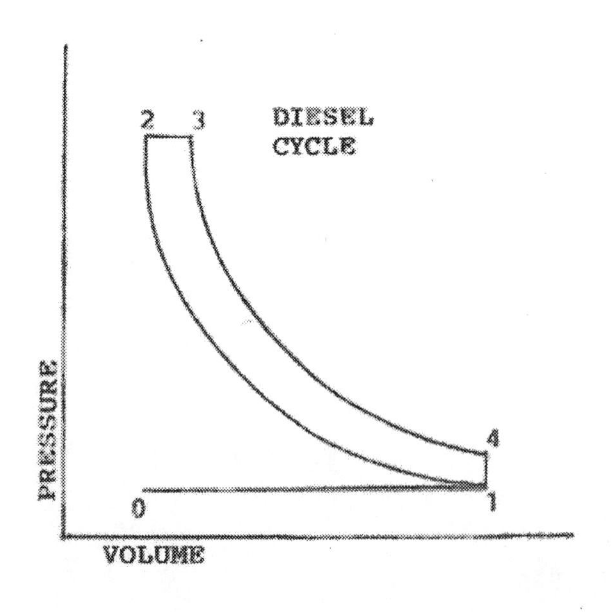
DIESEL
CYCLE
PRESSURE
VOLUME
2
3
4
0
1

Figure 25

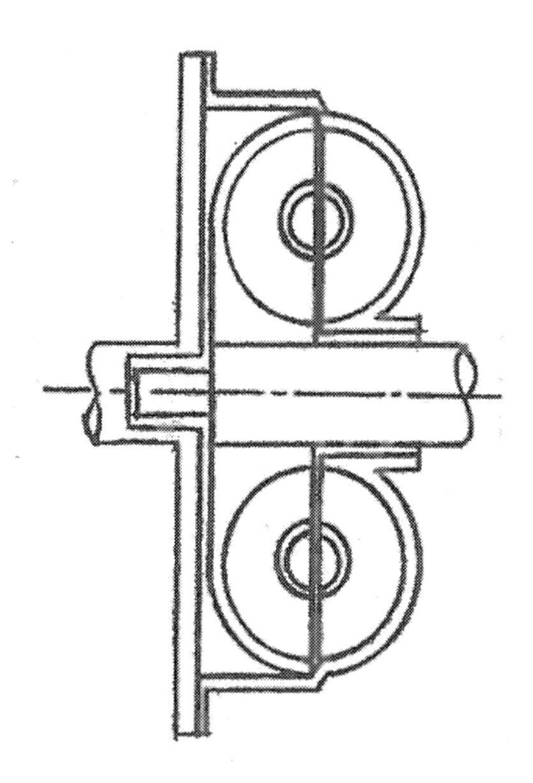

Torque Converter

torus where another set of blades called the "runner" is located. The oil is then recirculated inward back to the impeller where the cycle is repeated. In the fluid coupling, power is transmitted not only by mechanical means but by oil flowing across the narrow gap between the two rotors. This feature makes fluid couplings an excellent isolator of engine shock and vibration and negates the need for a manually operated clutch in automotive vehicles.

Passenger cars may have an automatic transmission or a manual transmission. Transmissions deliver engine power from the torque converter to the wheels. Automatic transmissions employ planetary gearsets to deliver power to the wheels in a number of different steps (ratios) forward and one in reverse. (See Figure 26.) The lowest or "first gear" is obtained by driving the "sun" gear, locking the "carrier" with the "ring gear" being the driven member. Other ratios and reverse operation can be obtained by using other combinations of the same gears. Front drive vehicles employ a "transaxle" which is a combination of the automatic transmission and the drive axle to deliver power from the front engine to the front wheels.

Figure 27 has a schematic of an automotive three-speed manual transmission. It incorporates an input shaft, a counter shaft and an output shaft. The input shaft rotates the countershaft which rotates the output shaft either through the first set of gears or the second set of gears. Third gear is obtained when the synchronizing clutch is engaged and power is transmitted from the input shaft directly through to the output shaft. Figure 28 has a step-by-step explanation of the power flow for the three different shift positions. Figure 29 has schematics explaining the operation of the synchronizing clutch. Figure 30 has a schematic of a Rzeppa joint. Rzeppa joints are used to connect the transmission to the wheels of a front drive automotive vehicle. Rzeppa joints are constant velocity couplings and can operate under high misalignments allowing ample wheel turning radius for steering. Figure 31 has a drawing of an "integral spindle wheel bearing". It is used on front drive, front steering vehicles. It has a female spline into which the Rzeppa joint male spline assembles into. This unit bolts to the vehicle to the right and the wheels bolt to it on the left. It is pre-lubricated and sealed for life. Figure 32 has a similar unit without the spline for non-drive wheels.

Figure 26

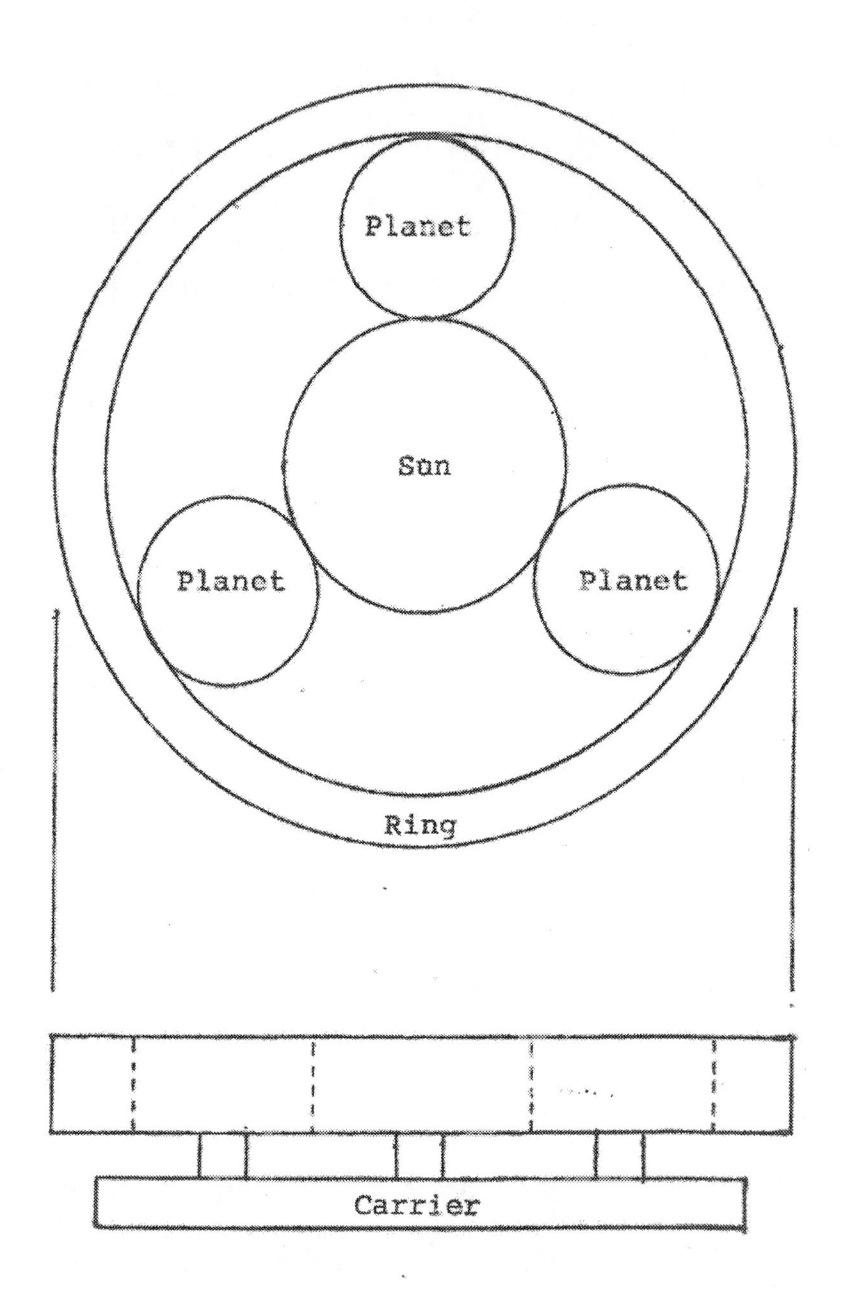

Planetary Gear Set

Figure 27

Automotive Manual Transmission Schematic

(Shown in Neutral Position)

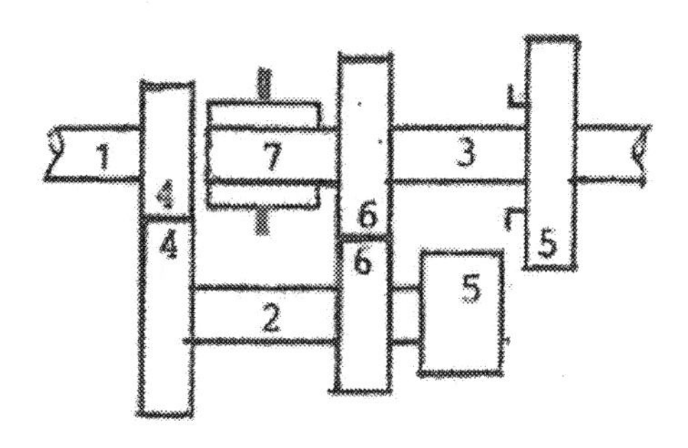

No 1 is the input shaft (driven by engine).

No 2 is the counter shaft.

No 3 is the output shaft (powers vehicle).

No 4 are counter shaft drive gears (always engaged).

No 5 are first gears (upper slides on shaft).

No 6 are second gears (upper spins freely).

No 7 is the syncronizing clutch (neutral position).

Figure 28

Automotive Manual Transmission Shift Positions

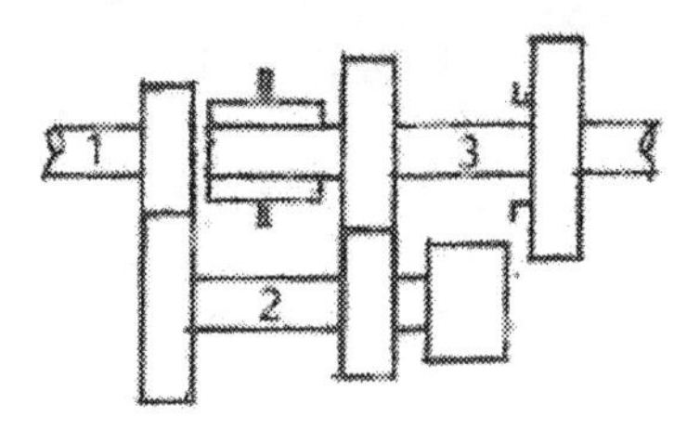

Neutral:
Input shaft(1) drives counter shaft(2) which spins freely. No power delivered to output shaft(3).

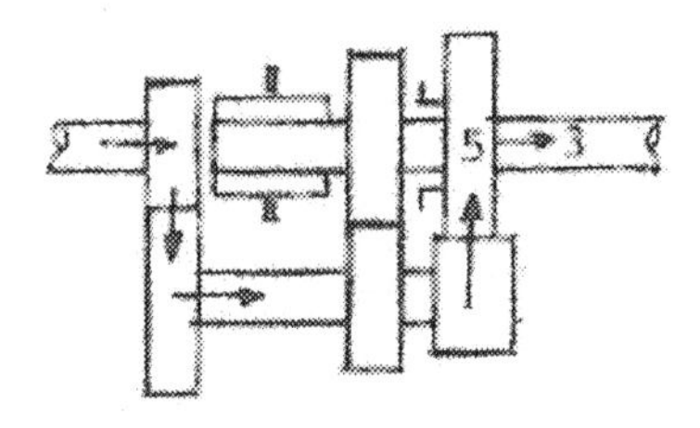

First:
Upper first gear(5) slid to the left engaging the output shaft(3).

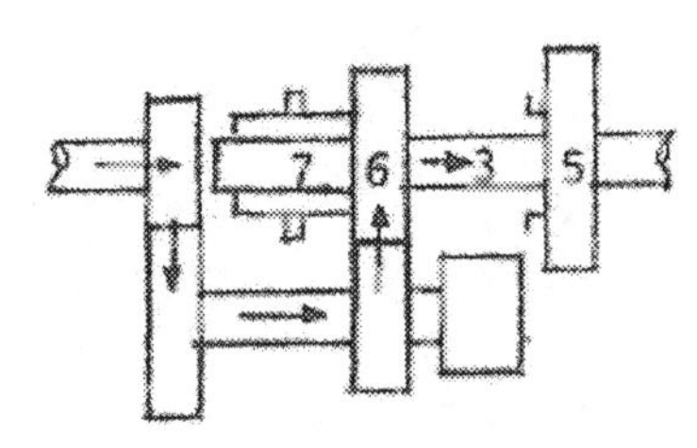

Second:
Upper first gear(5) is disengaged. The syncronizing clutch(7) locks second gear(6) to the output shaft(3).

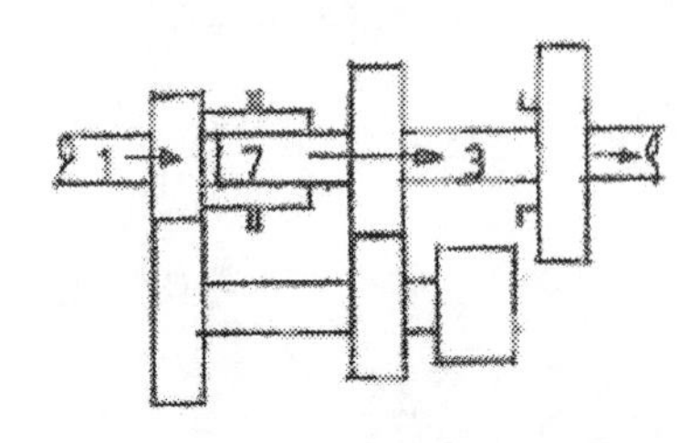

Third:
The syncronizing clutch(7) connects the input shaft(1) to the output shaft(3) resulting in direct drive.

Figure 29

Manual Transmission Synchronizer

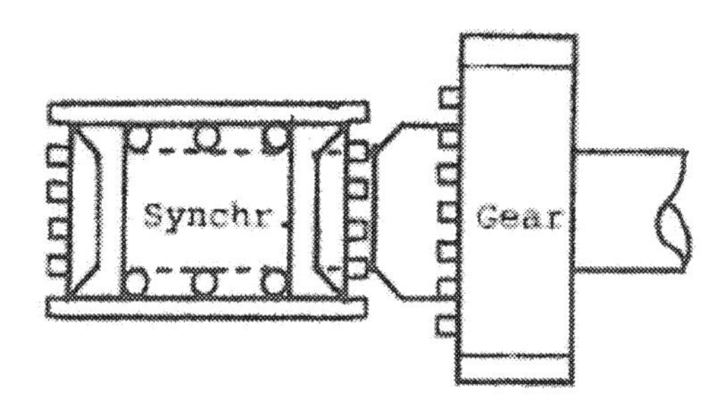

Synchronizer rotates with shaft.
Gear free to spin on same shaft.

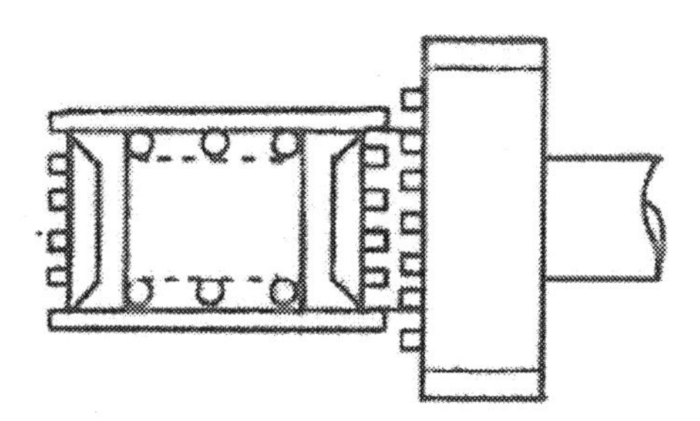

Synchronizer moved to the right.
Conical surfaces contact rotating
synchronizer and gear as a unit.

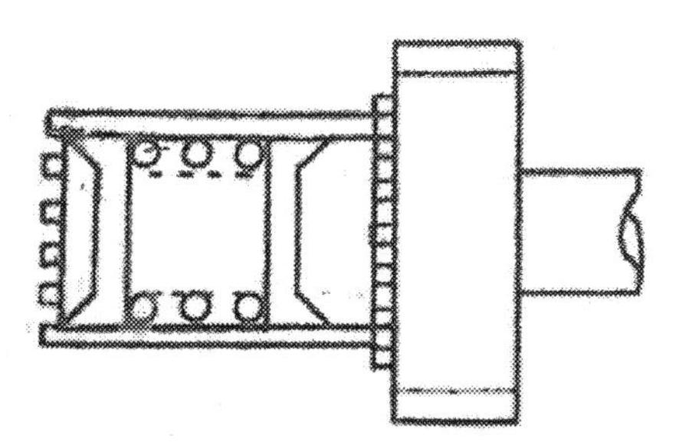

Synchronizer moved further to the
right until teeth positively engage.

Figure 30

Rzeppa Universal Joint

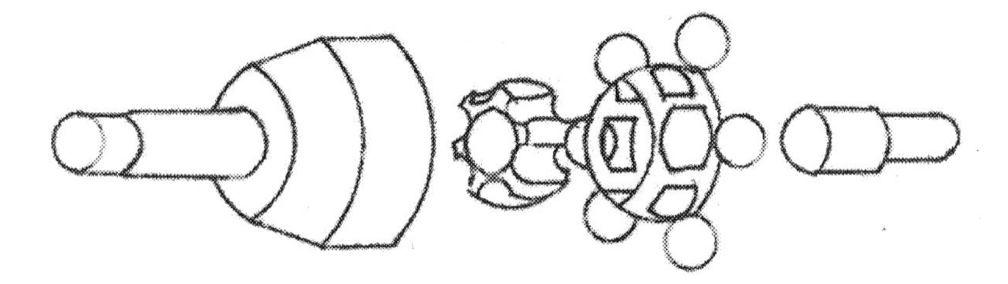

Figure 31

Integral Spindle

Drive Wheel Bearing

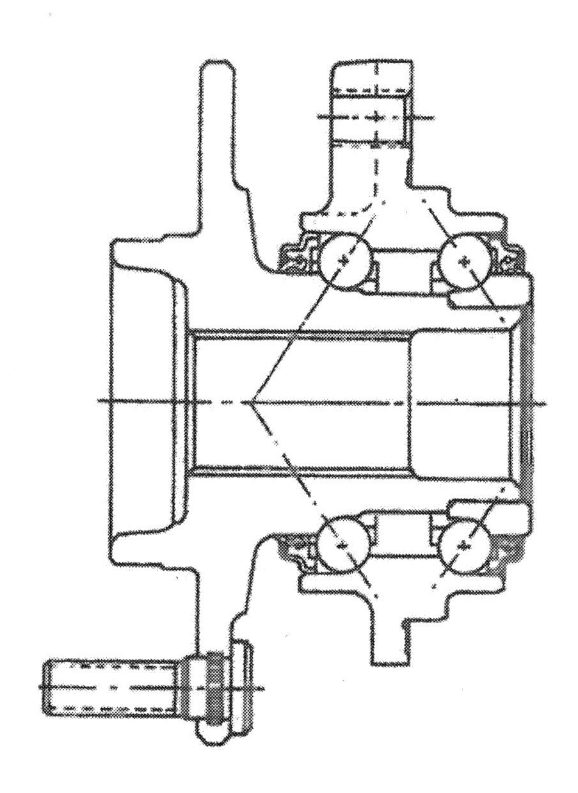

Figure 32

Integral Spindle

Non-Drive Wheel Bearing

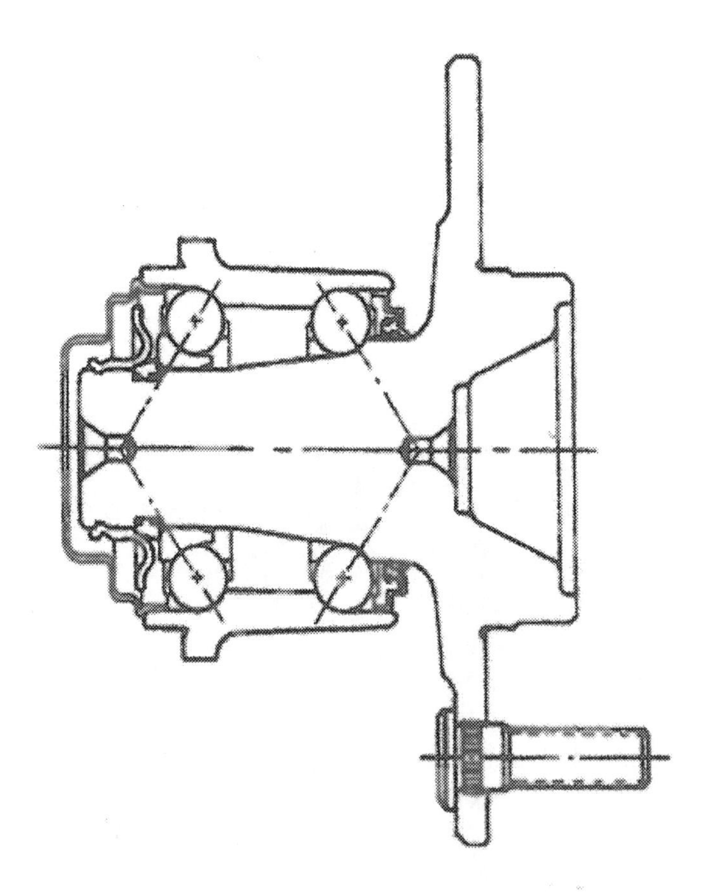

ENGINEERING PATENTS

There are three types of patents: "utility" patents, "design" patents and "plant" patents. Engineering patents usually fall into the category of utility patents. Utility patents involve materials, machines, components, and manufacturing parts and processes. Design patents involve only the appearance of an article while plant patents, as the name suggests, are granted to those that reproduce a new variety of plant.

When a new idea that is worth applying a patent for is conceived, a drawing or sketch should be made and a description written that makes the idea understandable to people working in the same field. If applicable, the drawing and written description should explain what is currently used, what its deficiencies are, and how they can be remedied by incorporating the new invention. The information should be prepared in a detailed manner and should include one or more claims made by the inventor for his invention. (Under current law, the patent effective date is when the idea is first thought of. Under revisions recently passed by the U.S. Congress, the patent effective date will be the actual filing date with the U.S. Patent Office.)

After all the preliminary work has been done, a Patent Attorney should be consulted. The Patent Attorney, for a fee, will make a search for "prior art" with the U.S. Patent Office. Prior art includes any patent that has been written (or is pending) that is the same as or similar to the one being applied for. Assuming that the prior art shows that the proposed patent is a new and novel idea, the Patent Attorney will prepare the patent application and submit it to the U.S. Patent Office who will examine the document and decide if a patent should be granted. The government fee to apply for patents is more for larger corporations than it is for individuals. If a patent is granted, there is a nominal yearly maintenance fee. Currently there are more than 150,000 patents issued every year and there have been a total of 8 million patents issued by the U.S. Patent Office since its inception. Figure 33 is the first page of a patent granted to a major U.S. car company. The patent was one of several ideas proposed in an attempt to improve the performance of automotive coolant pumps (water pumps) which had become a burden on vehicle warranty costs to the company.

Figure 33

United States Patent [19]

[11] **Patent Number:** **4,645,432**

[45] **Date of Patent:** **Feb. 24, 1987**

[54] **MAGNETIC DRIVE VEHICLE COOLANT PUMP**

[75] Inventor:

[73] Assignee:

[21] Appl. No.: 829,305

[22] Filed: Feb. 14, 1986

[51] Int. Cl.⁴ F04B 17/00; F04B 35/04

[52] U.S. Cl. 417/420; 415/10

[58] Field of Search .... 417/420, 423 R, 362; 415/10, 122 R; 416/3; 310/104

[56] References Cited

U.S. PATENT DOCUMENTS

2,033,377	3/1936	Hunter	417/420
2,471,753	5/1949	Johnston	417/420
2,827,856	3/1958	Zozulin	417/420
2,939,974	6/1960	Knight	417/420
3,458,122	7/1969	Andriussi	416/3
3,627,445	12/1971	Andriussi	416/3
3,723,029	3/1973	Laing	417/420
3,732,445	5/1973	Laing	417/420
4,184,090	1/1980	Taiani et al.	417/420

Primary Examiner—Carlton R. Croyle
Assistant Examiner—Timothy S. Thorpe
Attorney, Agent, or Firm—Patrick M. Griffin

[57] **ABSTRACT**

A magnetic drive pump for use as a vehicle coolant pump. A fluid housing fixed to the engine block as an impeller mounted on the outside of a cylindrical support integrally stamped into a front wall of the housing. A pulley has a central hub that is rotatably mounted within the cylindrical support, coaxial with the impeller bearing. A web of the pulley and the impeller both face the housing front wall in closely spaced, parallel relation, with opposed matching magnetic drive elements. The structure is particularly simple and compact, and needs no cartridge or bearing seal.

3 Claims, 1 Drawing Figure

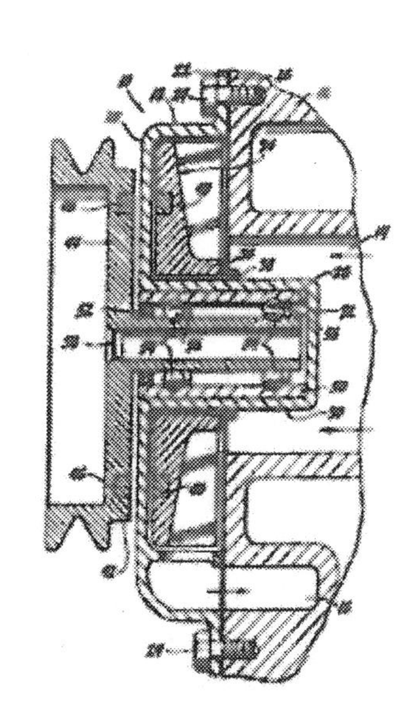

QUIZ

1) Iron ore is loaded in furnaces:

   a) With coke and limestone
   b) Blasted with hot air
   c) To produce pig iron
   d) All of the above

2) Chromium and nickel are added to steel largely to:

   a) Increase hardness
   b) Increase color
   c) Increase resistance to corrosion
   d) Increase ductility

3) Aluminum is used in industry because:

   a) It is lighter than steel
   b) It is easier cast than steel

c) Has more corrosion resistance than steel
d) All of the above

4) Magnesium is:

a) Lighter than aluminum
b) Heavier than aluminum
c) Obtained from mines
d) Obtained from lake water

5) The modulus of elasticity is a measure of the:

a) Strength of a material
b) Stiffness of a material
c) Hardness of a material
d) None of the above

6) The Brinell machine tests the:

a) Brittleness of a metal
b) Elasticity of a metal
c) Hardness of a metal
d) All of the above

7) Geometric Dimensioning and Tolerancing:

a) Is used for preparing engineering drawings
b) Does not replace conventional dimensioning
c) More precisely controls the fit between mating parts
d) All of the above

8) Geometric Dimensioning and Tolerancing:

a) Establishes a square as the theoretical boundary
b) Establishes a circle as the theoretical boundary
c) Establishes a rectangle as the theoretical boundary
d) None of the above

9) A Number 1 screw thread fastener tolerance is for:

a) Easy assembly
b) General assembly
c) Accurate assembly
d) Precise assembly

10) For shafts with a high amount of misalignment a:

a) Rigid coupling is used
b) Flexible coupling is used

c) Universal joint is used
d) None of the above

11) Belts are:

a) Used when shafts is too far apart for gears
b) A quiet and efficient means of transmitting power
c) Able to absorb vibration
d) All of the above

12) Poly V belts:

a) Are not as strong as V belts
b) Are more flexible than V belts
c) Do not last as long as V belts
d) All of the above

13) A beam formula can be used to calculate:

a) Stress
b) Deflection
c) Load
d) All of the above

14) Shafting design formula consider:

a) Torsion
b) Bending
c) Axial loading
d) All of the above

15) Anti-friction bearings operate in the high:

a) 90% range
b) 80% range
c) 70% range
d) 60% range

16) Anti-friction bearing life increases with:

a) Decreasing bearing capacity
b) Increasing bearing load
c) Decreasing bearing speed
d) Increasing bearing speed

17) Recommended bearing oil viscosity increases with:

a) Increasing speed
b) Higher operating temperature

c) Increasing bearing bore
d) All of the above

18) Thru-boring a housing allows for:

a) Better gear alignment
b) Better bearing alignment
c) Easier removal of the pinion gear
d) All of the above

19) Automotive center section differential gears:

a) Allow for either wheel to rotate faster than the other
b) Use planetary gears to control wheel speed
c) Use spur gears to control wheel speed
d) None of the above

20) Most gears operate in the high:

a) 60% range
b) 70% range
c) 80% range
d) 90% range

21) A gear driven by a smaller gear 3/4 its own size:

a) Will rotate at 4/3 the speed of the smaller gear
b) Will deliver 3/4 the torque of the smaller gear
c) Will change the direction of rotation of the driven gear
d) All of the above

22) Diametral pitch is a measure of:

a) Tooth strength
b) Tooth size
c) Tooth width
d) Tooth height

23) Helical gears

a) Run smoother and quieter than spur gears
b) Support lower loads than spur gears
c) Run at a higher efficiency than spur gears
d) Impose only thrust loads on a shaft

24) Wilford Lewis:

a) Calculated gear tooth contact stress

b) Calculated tooth torsional stress
c) Found tooth stress greatest at the tip
d) None of the above

25) Heinrich Hertz:

a) Determined contact pattern between two bodies
b) Determined stress between two contacting bodies
c) Work is used in rating gear pitting resistance
d) All of the above

26) The overall ratio of a simple gear train:

a) Is the quotient of the individual ratios
b) Is the sum of the of the individual ratios
c) Is the product of the individual ratios
d) is the sum of the first and last ratio

27) Compound gear trains:

a) Have each gear on a separate shaft
b) Have multiple gears on one shaft
c) Occupy more linear space than simple gear trains
d) None of the above

28) Industrial gearboxes can deliver power:

a) From a high speed source to a low speed device
b) From a motor to an end user
c) Using helical or worm gears
d) All of the above

29) Diesel engines:

a) Are generally used for heavy duty vehicles
b) Are general used for passenger cars
c) Use spark ignition
d) Operate with lower cylinder pressures

30) Torque converters:

a) Are mounted behind the engine and transmission
b) Transmit torque through a film of oil
c) Have a flywheel and impeller
d) All of the above

31) Automatic transmissions:

a) Deliver power from the torque converter to the wheels

b) Use simple gear trains to deliver power
c) Use planetary gears to deliver power
d) None of the above

32) The synchronizer delivers power to:

a) First gear
b) The output shaft
c) Second gear
d) Third gear

33) An engineering patent:

a) Is a utility patent
b) Is a design patent
c) Is a plant patent
d) None of the above

ANSWER KEY

1) d
2) c
3) d
4) a
5) b
6) c
7) d
8) b
9) a
10) c
11) d
12) b
13) d
14) d
15) a
16) c
17) b
18) d
19) a
20) d
21) c
22) b
23) a
24) d
25) d
26) c
27) b
28) d
29) a
30) d
31) c
32) b
33) a

LEARNING OBJECTIVES

This course teaches the following specific knowledge and skills:

- Improvements in engineering drawing methods using Geometric Dimensioning and Tolerancing.
- The advantage of using Poly V belts over standard V belts in automotive applications.
- How beam formula can be used to analyse shaft and bearing conditions in gear boxes.
- The use of important formula in the design of gears, bearings, and shafts.
- The difference between simple and compound gear trains.
- The difference between the Otto and diesel automotive engine cycles.
- The function of an automotive torque converter and the power flow through it.
- The power flow through the various gears of an automotive manual transmission.

- The operation and function of an automotive manual transmission synchronizer.
- When an idea is patentable and how to obtain an engineering patent.

ABOUT THE AUTHOR

The author has a BSME from Case-Western University, Cleveland, Ohio. He is a registered Professional Engineer in the State of Ohio. He has had 40 years of Mechanical Engineering experience, 26 of which were with the General Motors Corporation. While there, he obtained U.S. Patent number 4,645,432, “Magnetic Drive Vehicle Coolant Pump”. He went on to become a leader in anti-friction bearing applications in both the automotive and industrial fields. Valuable experience was also gained in gears and mechanical power transmission. Prior to that he was employed by TRW, Cleveland, Ohio, where he was responsible for bearings, gears and mechanical power transmission in the aircraft and missile fields under the tutelage of Mr. Thomas Barish, a leading mechanical power transmission consultant. Also, Mr. Tata has authored 25 technical papers that are available on the internet and other sources for professional development hours. He is also the author of the book “The Development of U.S. Missiles During the Space

Race with the U.S.S.R.". It is based on his experience, early in his career, working as a Flight Test Engineer at Cape Canaveral, Florida during the Cold War with the U.S.S.R. More recently, Mr Tata has ventured outside the technical field in authoring his second book, "The Greatest American Presidents". Following that is his third work, a part technical, part historical book titled "How Detroit became the "Automotive Capital of the World".